The Urban Book Series

The Urban Book Series is a resource for urban studies and geography research worldwide. It provides a unique and innovative resource for the latest developments in the field, nurturing a comprehensive and encompassing publication venue for urban studies, urban geography, planning and regional development.

The series publishes peer-reviewed volumes related to urbanization, sustainability, urban environments, sustainable urbanism, governance, globalization, urban and sustainable development, spatial and area studies, urban management, transport systems, urban infrastructure, urban dynamics, green cities and urban landscapes. It also invites research which documents urbanization processes and urban dynamics on a national, regional and local level, welcoming case studies, as well as comparative and applied research.

The series will appeal to urbanists, geographers, planners, engineers, architects, policy makers, and to all of those interested in a wide-ranging overview of contemporary urban studies and innovations in the field. It accepts monographs, edited volumes and textbooks.

Indexed by Scopus.

Jelena J. Stanković · Alessandro Luè ·
Claudia van der Laag
Editors

Regional and Urban Challenges and Opportunities for Data-Policy Interactions

The Western Balkans and Eastern European Countries

Springer

Editors
Jelena J. Stanković
Faculty of Economics
University of Niš
Niš, Serbia

Alessandro Luè
Poliedra—Politecnico di Milano
Milan, Italy

Claudia van der Laag
Department of Built Environment
Oslo Metropolitan University
Oslo, Norway

ISSN 2365-757X ISSN 2365-7588 (electronic)
The Urban Book Series
ISBN 978-3-032-08591-7 ISBN 978-3-032-08592-4 (eBook)
https://doi.org/10.1007/978-3-032-08592-4

This work was supported by the European Union under Grant Agreement number 101059994—UR-DATA—HORIZON-WIDERA-2021-ACCESS-02.

This Springer imprint is published by the registered company Springer Nature Switzerland AG
The registered company address is: Gewerbestrasse 11, 6330 Cham, Switzerland

Acknowledgments

The publication of this book is supported by the European Union under Grant Agreement number 101059994—UR-DATA—HORIZON-WIDERA-2021-ACCESS-02.

We would like to thank all the authors for their valuable contributions to this book.

Also, we greatly appreciated the comments from colleagues who served as reviewers for the chapters: Prof. Ivana Bilic, University of Split, Croatia; Dr. Isın Can Traunmüller, Izmir Institute of Technology, Turkey; Prof. Marija Džunić, University of Niš, Serbia; Dr. Aretoulis George, Aristotle University, Greece; Dr. Devon McAslan, Chalmers University of Technology, Sweden; Dr. Vinko Muštra, University of Split, Croatia; Prof. Milijana Novovic Buric, University of Montenegro, Montenegro; Prof. Iason Papathanasiou, University of Macedonia, Macedonia; Dr. Primož Pevcin, University of Ljubljana, Slovenia; Dr. Meri Šuman Tolić, University of Dubrovnik, Croatia; Dr. Marina Toger, Uppsala University, Sweden; Dr. Umut Turk, Abdullah Gül University, Turkey; and Dr. Andriela Vitic Cetkovic, University of Montenegro, Montenegro.

Praise for *Regional and Urban Challenges and Opportunities for Data-Policy Interactions*

"This volume presents a wealth of theories, methods, and empirical research on urban and regional development and on economies in the Balkan countries, with a specific focus on Serbia and the city of Niš. The authors apply cutting-edge research methods to current problems, ranging from street-level accessibility, walkability, and bikeability, to urban planning, design, and perspectives of tourism and heritage, up to the regional level of economic, environmental, and social development. While many chapters refer to regions or cities in Serbia and the Balkan region, the underlying trends of urban and regional development, as well as the theoretical underpinnings explaining these, can be found in numerous countries around Europe and worldwide. This book is not only recommended to readers interested in current topics of development economics applied to transition regions, especially in the Balkan countries. It is also a good read for a general audience eager to learn about the application of a large diversity of economic theories and methods—ranging from highly qualitative to quantitative research—in an interesting and challenging regional context."

—Prof. Michael Getzner, *TU Wien, Vienna, Austria*

Contents

1 **Introduction** 1
Jelena J. Stanković, Alessandro Luè, and Claudia van der Laag

Part I Demographics

2 **Uncovering the Drivers of Urbanization in Serbia: An Econometric Investigation** 13
Jelena J. Stanković, Marija Džunić, Srđan Marinković, and Ivana Marjanović

3 **Exploring the Relationship Between Economic and Demographic Performance of Local Governments: An Analysis of Factors and Implications** 33
Marina Čolig, Kristina Kregar, and Saša Drezgić

4 **Urban Development Paths: Understanding the Intra-Regional Differentiation in the Region of Southern and Eastern Serbia** 61
Marija Antić, Dejan Ž. Đorđević, and Danica Đurkin

5 **Is Return Migration the End of the Migration Cycle? Evidence from Serbia** 85
Milica Langović, Danica Šantić, and Dejan Ž. Đorđević

Part II Overcoming Inequality: Governance and Institutions

6 **Efficiency and Performance of Sustainable Development of the Western Balkan Countries: Whether Membership in the European Union is Recipe for Sustainability?** 103
Predrag Mimović, Ana Krstić Srejović, and Milena Jakšić

7 **Smart Participation for Smarter Cities: Smart Citizens' Capacity Framework** 131
Ivana Simić

8 Key Determinants of the Quality of Towns and City Centres in the Function of More Resilient and Sustainable Tourism Development Management 153
Sonja Jovanović, Vesna Janković Milić, and Vinko Lepojević

Part III Smart Growth in Cities

9 On Foot Through Town? The Conundrum of Pedestrian Accessibility in Niš, Serbia 173
Claudia van der Laag, Wendy Tan, Chiara Garau, and Beniamino Murgante

10 Urban Bikability: A Case-Comparison of Utrecht, the Netherlands and Nis, Serbia 189
Daniel Piatkowski, Chaoru Lu, and John Östh

11 Placemaking of Urban Green Areas: The Failed Potential of Social Interaction 205
Tanja Obradović, Jelena Đekić, Tamara Rađenović, Milena Dinić Branković, and Milica Igić

12 Enhancing Natural and Cultural Heritage Through Sustainable Tourism and Improved Accessibility to Support Economic Development of Sensitive Regions in the Balkan and Eastern European Countries 239
Ilenia Pierantoni, Nora Annesi, and Massimo Sargolini

Author Index 255

Editors and Contributors

About the Editors

Jelena J. Stanković is Full Professor at the Faculty of Economics, University of Niš, Serbia in the scientific field of mathematics and statistics in economics. Her research focuses on quantitative economic analysis, especially in multi-criteria decision-making for local economic development issues, urban labor market, and smart and sustainable development of urban areas. She is Active Member of the Committee for Economic Sciences of the Serbian Academy of Science and Arts. She has been involved in a number of national and international projects as Researcher and Principal Investigator. Professor Stanković is Editorial Board Member of the journal *Review: Journal of Social Issues*, published by the University of Sarajevo, the *Journal of Economics and Business* published by the University of Rijeka, and the *International Journal of Sustainable Agricultural Management and Informatics* published in Switzerland.

Alessandro Luè is Managing Director and Head of Research of Poliedra, the consultancy and service center at Politecnico di Milano, Italy. Further, he is Adjunct Professor of Methods and Models for Transport Decisions at Politecnico di Milano and has over 20 years of experience in applied research. His research interests include decision support systems, multi-criteria analysis, sharing mobility, electric mobility, mobility management, intelligent transport system, smart cities and communities. Dr. Luè worked as Research Associate at the Intelligent Transportation Lab of MIT—Massachusetts Institute of Technology, USA and received in 2015 his Ph.D. in Design from Politecnico di Milano. He has been involved in numerous national and international research and consultancy projects.

Claudia van der Laag is Full Professor of Urban Analytics in the Department of Built Environment at OsloMet—Oslo Metropolitan University, Norway. In addition, she serves as Senior Expert to the Urban Development Director's Office of the City of Linz, Austria and is Norman Foster Foundation Academic Associate. Her expertise on transport and land use planning including people's spatial behavior combines a wide range of analytical techniques with mixed methods at the forefront of virtual modeling at the intersection of urban planning and urban design. Professor van der Laag implements and develops new methods and tools and shapes policies for sustainable cities and regions. She combines theoretical innovation with practice-oriented solutions and has been involved in numerous international projects. She is Co-author of the textbook "Introduction to Space Syntax in Urban Studies," Editorial Board Member of Springer's The Urban Book Series and Associate Editor of the Taylor and Francis journal "Sustainable Communities." In 2015 she received the prestigious Michael Breheny Prize for her work on complexity-based planning models.

Contributors

Nora Annesi Department of Business, Law, Economics and Consumer Behaviour, IULM University, Milan, Italy

Marija Antić Faculty of Geography, University of Belgrade, Beograd, Serbia

Milena Dinić Branković Faculty of Civil Engineering and Architecture, University of Niš, Niš, Serbia

Marina Čolig Municipality of Zagreb, Croatia

Jelena Đekić Faculty of Civil Engineering and Architecture, University of Niš, Niš, Serbia

Dejan Ž. Đorđević Faculty of Economy, University of Niš, Niš, Serbia

Saša Drezgić Faculty of Economics, University of Rijeka, Rijeka, Croatia

Danica Đurkin Faculty of Geography, University of Belgrade, Beograd, Serbia

Marija Džunić Faculty of Economics, University of Niš, Niš, Serbia

Chiara Garau DICAAR, University of Cagliari, Cagliari, Italy

Milica Igić Faculty of Civil Engineering and Architecture, University of Niš, Niš, Serbia

Milena Jakšić Faculty of Economics, University of Kragujevac, Kragujevac, Serbia

Sonja Jovanović Faculty of Economics, University of Niš, Niš, Serbia

Kristina Kregar University of Rijeka, Rijeka, Croatia;
University of Trieste, Trieste, Italy

Ana Krstić Srejović Faculty of Economics, University of Belgrade, Belgrade, Serbia

Milica Langović Institute of Physics Belgrade, University of Belgrade, Belgrade, Serbia

Vinko Lepojević Faculty of Economics, University of Niš, Niš, Serbia

Chaoru Lu Department of Built Environment, Oslo Metropolitan University, Oslo, Norway

Alessandro Luè PoliedraPolitecnico Di Milano, Milan, Italy

Srđan Marinković Faculty of Economics, University of Niš, Niš, Serbia

Ivana Marjanović Faculty of Economics, University of Niš, Niš, Serbia

Vesna Janković Milić Faculty of Economics, University of Niš, Niš, Serbia

Predrag Mimović Faculty of Economics, University of Kragujevac, Kragujevac, Serbia

Beniamino Murgante School of Engineering, University of Basilicata, Potenza, Italy

Tanja Obradović Department for Inspection Affairs of Civil Engineering and Urban Planning, Ministry of Construction, Transport and Infrastructure, Niš, Serbia

John Östh Department of Built Environment, Oslo Metropolitan University, Oslo, Norway

Daniel Piatkowski Department of Built Environment, Oslo Metropolitan University, Oslo, Norway

Ilenia Pierantoni School of Architecture and Design, University of Camerino, Camerino, Italy

Tamara Rađenović Faculty of Occupational Safety, University of Niš, Niš, Serbia

Danica Šantić Faculty of Geography, University of Belgrade, Belgrade, Serbia

Massimo Sargolini School of Architecture and Design, University of Camerino, Camerino, Italy

Ivana Simić University of Niš, Niš, Serbia

Jelena J. Stanković Faculty of Economics, University of Niš, Niš, Serbia

Wendy Tan Wageningen University Research, Wageningen, The Netherlands; Western Norway University, Bergen, Norway

Claudia van der Laag Faculty of Technology, Art and Design, Oslo Metropolitan University, Oslo, Norway;
Department of Built Environment, Oslo Metropolitan University, Oslo, Norway

Chapter 1
Introduction

Jelena J. Stanković, Alessandro Luè, and Claudia van der Laag

Abstract The Balkan and Eastern European regions are at a critical juncture in their urbanization and regional development trajectories. The United Nations (https://unhabitat.org/sites/default/files/2022/06/wcr_2022.pdf) predicts that by 2050, 68% of the world's population will live in urban areas, a trend that is expected to be even more pronounced in developing regions, including the Balkans. Urbanization, the movement of populations from rural to urban areas, has long been recognized as a key driver of economic growth and modernization (Guo et al. in Environ Dev Sustain 24(5):6508–6526, 2022). However, as global and regional trends in urbanization continue to reshape cities and towns, the challenges of managing demographic changes, economic disparities, and environmental sustainability become more pressing (Fratesi in Regional Policy: Theory and Practice. Taylor and Francis, London, 2023).

Keywords Balkan and Eastern Europe region · Employment rates · Transportation infrastructure · Crime levels · Cultural amenities

1.1 Background

The Balkan and Eastern European regions are at a critical juncture in their urbanization and regional development trajectories. The United Nations (2022) predicts that by 2050, 68% of the world's population will live in urban areas, a trend that is expected to be even more pronounced in developing regions, including the Balkans.

J. J. Stanković (✉)
Faculty of Economics, University of Niš, Niš, Serbia
e-mail: jelena.stankovic@eknfak.ni.ac.rs

A. Luè
PoliedraPolitecnico Di Milano, Milan, Italy

C. van der Laag
Faculty of Technology, Art and Design, Oslo Metropolitan University, Oslo, Norway

J. J. Stanković et al. (eds.), *Regional and Urban Challenges and Opportunities for Data-Policy Interactions*, The Urban Book Series,
https://doi.org/10.1007/978-3-032-08592-4_1

Urbanization, the movement of populations from rural to urban areas, has long been recognized as a key driver of economic growth and modernization (Guo et al. 2022). However, as global and regional trends in urbanization continue to reshape cities and towns, the challenges of managing demographic changes, economic disparities, and environmental sustainability become more pressing (Fratesi 2023).

In the context of the Balkans and Eastern Europe, the process of urbanization is intricately linked to the region's unique historical, political, and socio-economic conditions. Urbanization in these regions accelerated significantly after the transition from centrally planned economies to market-driven systems in the early 1990s. The collapse of socialist regimes and the introduction of market economies opened up new avenues for economic growth, but also exposed these countries to new vulnerabilities, particularly in terms of governance, infrastructure, and spatial planning (Vasilevska et al. 2020). Economically, the region struggles with weak market integration, limited absorption of EU financial assistance, and slow progress in building a competitive regional economy. The EU's response, shaped by internal crises such as Brexit, financial instability, and the COVID-19 pandemic, has been constrained, delaying accession processes and reducing financial commitments (Qorraj et al. 2024). Additionally, the Western Balkans exhibit features of the middle-income trap, marked by insufficient innovation capacity, low productivity, and limited competitiveness on global markets, further exacerbating socio-economic stagnation (Muštra & Rodríguez-Pose 2021).

The transition led to the rise of urban centers as hubs for economic activity, migration, and innovation, but also created significant urban–rural disparities and environmental pressures (Glaeser et al. 2014). The emergence of major cities, such as Belgrade, Skopje, and Sofia, as economic powerhouses, attracting investment and skilled labor, contrasts sharply with the stagnation or decline of smaller towns and rural areas (Stanković et al. 2021). The uneven development has resulted in significant disparities in access to services, infrastructure, and employment opportunities, thus enhancing the urban–rural divide. Studies have referred to this phenomenon as "urban primacy" in the Balkans, where a single city dominates the national urban system in terms of size, economic importance, and political influence (Tošić and Krunić 2005). The issue of regional disparities is further exacerbated by depopulation trends, particularly in rural and remote areas, where migration to urban centers and abroad has depleted the labor force and strained local economies (Đurkin et al. 2021). The rapid growth of cities in the Balkans and Eastern Europe has also put immense pressure on natural resources, infrastructure, and public services (Knez et al. 2022). Modern cities, in particular, face significant challenges due to high levels of resource consumption, pollution, and habitat loss driven by population density and associated economic activities (Ignjatović et al. 2024; Oktay 2024). The heavy dependence on private transportation and traffic congestion exacerbates air pollution and increases carbon emissions. Moreover, urban prosperity often brings issues such as housing affordability, displacement, gentrification, and growing social inequality. According to the findings of Belis et al. (2023), the health impacts and associated external costs linked to air pollution in 2019 were evaluated across 30 urban areas within the Western Balkans region, recognized as one of Europe's major air pollution hotspots.

Additionally, in the Western Balkans, the public may lack access to sufficient or accurate information about the main sources of air pollution, highlighting the need for more transparent communication and enhanced education on air pollution monitoring (Madjar et al. 2023).

Consequently, addressing these challenges requires a comprehensive regional development strategy that fosters sustainable urban planning and governance across the entire urban–rural spectrum, by balancing economic growth with environmental protection and social equity.

The European Union (EU) has been a significant player in shaping urban and regional policies in Eastern Europe, particularly through the Urban Agenda for the EU, which emphasizes sustainable and inclusive urban development (European Commission 2006). This agenda is aligned with the Sustainable Development Goals (SDGs), particularly SDG 11, which focuses on making cities inclusive, safe, resilient, and sustainable. The EU's Green Deal further highlights the importance of transitioning toward greener, more sustainable urban economies, with a focus on reducing carbon emissions, improving energy efficiency, and promoting the circular economy.

Despite the Balkan and Eastern Europe region been historically been characterized by weak institutions, corruption, and inefficiencies in public administration (Bartlett 2009), recent reforms aimed at improving governance, transparency, and citizen participation. The concept of "smart governance" has gained traction in the region, particularly in the context of the Smart City initiatives, which leverage digital technologies to improve urban services, enhance citizen engagement, and promote sustainability. Institutional reforms are essential for creating the conditions for sustainable urban growth. The capacity of local governments to manage urbanization processes effectively is often constrained by limited financial resources, inadequate infrastructure, and a lack of technical expertise. However, initiatives such as the European Structural and Investment Funds (ESIF) provide crucial financial support for infrastructure development, environmental protection, and social inclusion in the region (Bartlett 2009). These funds have been fundamental in supporting projects that improve urban mobility, enhance energy efficiency, and promote social cohesion in cities across the Balkans and Eastern Europe.

In addition to EU initiatives, national policies in the Western Balkans, such as Serbia's Sustainable Urban Development Strategy 2030, underscore the importance of addressing the environmental and social dimensions of urbanization. This strategy aims to promote balanced regional development, reduce inequalities, and enhance the quality of life in both urban and rural areas. A key component of this strategy is the emphasis on data-driven policymaking, where the collection and analysis of urban data informs decisions on infrastructure development, public services, and environmental protection (Stanković et al. 2021). Data plays a critical role in informing urban policies and planning decisions. In recent years, there has been a growing emphasis on the use of data analytics and econometric models to understand the drivers of urbanization and assess the effectiveness of policy interventions (Stanković et al. 2021). For example, the use of panel regression analysis in urban studies has enabled researchers to identify key factors such as employment, transportation, and cultural amenities that influence urban attractiveness and growth.

These insights are essential for policymakers seeking to develop evidence-based strategies for sustainable urban development. The interaction between data and policy is particularly important in the context of the Balkan region, where rapid urbanization and regional disparities require targeted interventions. Data-driven approaches can help local governments allocate resources more efficiently, monitor the impact of urban policies, and adjust strategies based on real-time feedback. Moreover, the integration of Geographic Information Systems (GIS) and other digital tools into urban planning processes allows for a more comprehensive analysis of spatial patterns, infrastructure needs, and environmental impacts.

This edited volume *"Urban and Regional Challenges and Opportunities in the Balkan and Eastern European Countries"* is produced as part of research funded by the European Union under grant agreement number 101059994—UR-DATA—HORIZON-WIDERA-2021-ACCESS-02. The views and opinions expressed in this publication are solely those of the author(s) and do not necessarily reflect the official position of the European Union or the European Research Executive Agency. The European Union and the European Research Executive Agency assume no responsibility for the content presented herein.

1.1.1 Overview of the Book's Structure and Chapter Summaries

This book is divided into three parts, each addressing key themes in urban and regional development in the Balkan and Eastern European countries. Each part of the book offers a diverse range of perspectives, drawing on empirical research, policy analysis, and theoretical frameworks.

In contemporary urban and regional planning, the interaction between data and policy has become an essential tool for guiding decision-making and addressing the complex challenges of urbanization. The chapters in this book not only provide valuable academic insights but also highlight the critical role of data-driven policy-making in shaping urban and regional development. Besides the content preview, this section provides an overview of how data-policy interactions are featured across the three parts of the book, with each chapter contributing with unique perspectives and analyses that directly inform urban policies.

PART 1: Demographics

The first part of the book focuses on demographic trends and urbanization patterns, employing data-driven approaches such as econometric models and regional analyses to inform policies that address the demographic challenges faced by cities and regions in the Balkans.

Chapter 2: Uncovering the Drivers of Urbanization in Serbia: An Econometric Investigation

This chapter by Stanković et al. highlights the critical role of panel regression analysis in understanding the drivers of urbanization in 24 Serbian cities over a decade. The authors utilize a wide range of indicators—such as employment rates, transportation infrastructure, crime levels, and cultural amenities—to construct an econometric model that identifies the key factors influencing urban attractiveness. This approach demonstrates the value of quantitative data in shaping policies for urban growth. For example, the finding that crime rates negatively affect urban appeal offers clear guidance for policymakers to prioritize public safety measures. The interaction between this empirical data and policy implications is evident in the recommendation that cities enhance cultural infrastructure and transport accessibility to foster urbanization. This model provides a valuable tool for regional authorities in Serbia to tailor urban policies that address both economic growth and social well-being.

Chapter 3: Exploring the Relationship Between Economic and Demographic Performance of Local Governments: An Analysis of Factors and Implications

Colig et al. employ the Tiebout theorem as a conceptual framework to analyze the relationship between economic and demographic performance in local governments. By examining how data on public goods provision, job creation, and demographic characteristics influence local economic growth, the chapter demonstrates how evidence-based policy can enhance local government performance. The Tiebout theorem's emphasis on "voting with one's feet" (where individuals choose jurisdictions that provide the services they prefer) underpins the need for local governments to make data-driven decisions. Policies informed by this research could focus on improving public service delivery, creating jobs, and providing better educational and healthcare services to retain and attract residents. Local governments can use such demographic and economic data to create targeted policies that enhance competitiveness and stimulate regional development.

Chapter 4: Urban Development Paths: Understanding the Intra-Regional Differentiation in Southern and Eastern Serbia

This chapter employs spatial analysis and regional typologies to explore the disparities in urban development across Southern and Eastern Serbia. By utilizing regional data, including economic performance indicators and demographic trends, the authors demonstrate the uneven spatial development within the region. The interaction between this data and policy is crucial: the identification of regional imbalances can guide policymakers in implementing targeted interventions aimed at fostering more balanced development. For example, regional governments might focus on improving infrastructure in less-developed areas or fostering economic diversification. The chapter calls for a data-driven regional strategy to address the socio-economic disparities within Serbia's regions, ensuring that urban policies are inclusive and equitable.

Chapter 5: Is Return Migration the End of the Migration Cycle? Evidence from Serbia

Todorović et al. use migration data to examine the phenomenon of return migration in Serbia, investigating whether returning migrants settle permanently or continue to relocate. The data-policy interaction here is focused on shaping national and regional migration policies. The findings highlight the need for policies that not only facilitate the reintegration of returning migrants but also address the economic conditions that might drive continued migration. Regional authorities can use migration data to craft strategies that provide employment opportunities and support services for returning migrants, ensuring that return migration contributes positively to regional development.

PART 2: Overcoming Inequality: Governance and Institutions

The second part of the book focuses on the role of governance and institutions in addressing inequalities and promoting sustainable urban development. This part of the book delves into governance and institutional frameworks, exploring how data-driven insights can enhance governance practices and reduce inequality in urban development. The chapters in this section examine how institutional reforms, governance structures, and public participation can contribute to more equitable urban growth.

Chapter 6: Efficiency and Performance of Sustainable Development of the Western Balkan Countries: Whether Membership in the European Union Is a Recipe for Sustainability?

Mimović et al. explore the relationship between EU membership and the performance of Western Balkan countries in achieving sustainable development. The chapter draws on extensive comparative data to assess whether EU accession has driven improvements in sustainability metrics, such as environmental performance, economic resilience, and social inclusivity. This data-policy interaction underscores the importance of aligning national policies with EU frameworks, such as the Urban Agenda for the EU and the European Green Deal, which prioritize sustainable urbanization. Policymakers in the Western Balkans can use this comparative data to align their national strategies with EU benchmarks, ensuring that urban development policies meet the standards of environmental protection, economic sustainability, and social inclusion. By identifying best practices from EU countries, this chapter offers valuable insights for policymakers in non-EU Balkan states to create policies that promote sustainability and resilience.

Chapter 7: Smart Participation for Smarter Cities: Smart Citizens' Capacity Framework

In this chapter, Simic presents a smart citizen participation framework that leverages digital data to enhance citizen engagement in urban governance. The interaction between data and policy is evident in the use of digital platforms and open data to facilitate public participation in decision-making processes. This framework enables cities to gather real-time feedback from citizens, which can be integrated into urban planning and policy decisions. The use of digital tools, such as online surveys, participatory budgeting platforms, and urban data dashboards, offers a practical way for local governments to make more informed decisions that reflect the needs and preferences of citizens. Policies informed by this data can promote more inclusive and democratic urban governance, ensuring that urban development is responsive to the voices of residents.

Chapter 8: Key Determinants of the Quality of Towns and City Centres: Toward More Resilient and Sustainable Tourism Development

Jovanović et al. employ urban quality indicators to assess the factors that contribute to the attractiveness of city centers for tourism and sustainable development. Using data on urban design, public spaces, and tourism infrastructure, the chapter demonstrates how data-driven insights can inform policies aimed at enhancing urban resilience. The interaction between data and policy is evident in the chapter's recommendations for improving urban quality through targeted investments in public spaces, cultural heritage, and sustainable tourism infrastructure. Policymakers can use these insights to develop urban strategies that boost tourism while ensuring that cities remain livable, resilient, and environmentally sustainable.

PART 3: Smart Growth in Cities

The third part of the book focuses on smart growth strategies for promoting sustainable urban development. It explores innovative approaches to urban planning, mobility, and the use of public spaces.

Chapter 9: On Foot Through Town? The Conundrum of Pedestrian Accessibility in Niš, Serbia

This chapter by Van der Laag Yamu et al. uses pedestrian accessibility data to assess the walkability of Niš, Serbia. The authors utilize GIS-based analysis and pedestrian flow data to identify areas where urban design can be improved to enhance walkability. The data-policy interaction here is crucial for informing urban mobility policies that prioritize pedestrian infrastructure. The chapter's findings suggest that policymakers should invest in pedestrian-friendly infrastructure, such as wider sidewalks, pedestrian zones, and improved public transport connectivity, to reduce car dependency and promote sustainable urban mobility. By integrating pedestrian data into urban planning processes, local governments can create policies that encourage more sustainable and livable urban environments.

Chapter 10: Urban Bikability: A Case-Comparison of Utrecht, The Netherlands, and Niš, Serbia

Piatkowski et al. provide a comparative analysis of bikability data from Utrecht and Niš, drawing on metrics such as cycling infrastructure, safety, and usage patterns. The chapter demonstrates how urban cycling data can inform policies aimed at promoting sustainable transportation. The interaction between data and policy is evident in the chapter's recommendations for improving cycling infrastructure in Niš, such as building dedicated bike lanes, improving road safety for cyclists, and promoting cycling as a viable alternative to car travel. Policymakers in both cities can use these insights to develop comprehensive urban cycling policies that support environmental sustainability, public health, and reduced traffic congestion.

Chapter 11: Placemaking of Urban Green Areas: The Failed Potential of Social Interaction

Obradović et al. use spatial and social data to analyze the placemaking potential of urban green areas, focusing on how poorly designed public spaces have hindered social interaction. The chapter emphasizes the need for data-driven placemaking strategies that prioritize inclusivity and community engagement. The interaction between data and policy is reflected in the recommendation to redesign urban green spaces based on the preferences and behaviors of residents, gathered through surveys, observational studies, and spatial analytics. Policymakers can use these insights to create green spaces that foster social interaction, community cohesion, and improved quality of life.

Chapter 12: Enhancing Natural and Cultural Heritage through Sustainable Tourism

Pierantoni et al. explore how sustainable tourism can be used to enhance the natural and cultural heritage of sensitive regions in the Balkans and Eastern Europe. The chapter highlights the potential for tourism to support economic development while preserving the region's unique cultural and environmental assets. The focus is on tourism and heritage data to assess how natural and cultural sites in the Balkans can be enhanced through sustainable tourism practices. The chapter demonstrates how data on visitor patterns, environmental impacts, and local economies can inform policies that balance tourism development with heritage preservation. The interaction between data and policy is critical in crafting strategies that promote sustainable tourism while ensuring that natural and cultural heritage is protected for future generations.

References

Bartlett W (2009) Economic development in the European super-periphery: evidence from the Western Balkans. Econ Ann 54(181):21–44

Beall J, Fox S (2009) Cities and development. Routledge

Belis CA, Matkovic V, Ballocci M, Jevtic M, Millo G, Mata E, Van Dingenen R (2023) Assessment of health impacts and costs attributable to air pollution in urban areas using two different approaches: a case study in the Western Balkans. Environ Int 182:108347

Bignami D, Colorni A, Lué A, Nocerino R, Rossi M, Savaresi S (2017) Electric vehicle sharing services for smarter cities. Springer

Busso M, Chauvin JP, Herrera N (2021) Rural-urban migration at high urbanization levels. Reg Sci Urban Econ 91:103658

Đurkin D, Antić M, Budović A (2021) Demographic and economic aspects of urban shrinkage in Serbia-typology and regional differentiation. Glas Srp Geogr Drustva 101(2):43–78

European Commission (2006) Cohesion policy and cities: the urban contribution to growth and jobs in the regions. Communication from the Commission to the Council and Parliament COM(2006) 385 final of 13 Jul 2006, Brussels. https://eur-lex.europa.eu/

Fratesi U (2023) Regional policy: theory and practice. Taylor and Francis, London

Glaeser EL, Ponzetto GA, Tobio K (2014) Cities, skills and regional change. Reg Stud 48(1):7–43

Grimmond SUE (2007) Urbanization and global environmental change: local effects of urban warming. Geogr J 173(1):83–88

Guo J, Yu Z, Ma Z, Xu D, Cao S (2022) What factors have driven urbanization in China? Environ Dev Sustain 24(5):6508–6526

Gursoy O, Kodaz M (2021) Urbanization and environmental problems in the Western Balkans and policy recommendations. In: Korkut H (ed) Studies in urbanization and local governance in Balkan countries. Dobra knjiga, Sarajevo, pp 15–38

Ignjatović D, Miletić N, Zeković B, Ćuković Ignjatović N (2024) Cost-benefit analysis of building energy refurbishment–case study of a municipal public building. Econ Them 62(1):45–66

Knez S, Štrbac S, Podbregar I (2022) Climate change in the Western Balkans and EU Green Deal: status, mitigation and challenges. Energy, Sustain Soc 12(1):1–14

Madjar A, Gjorshoska I, Prodanova J, Dedinec A, Kocarev L (2023) Western Balkan societies' awareness of air pollution. Estimations using natural language processing techniques. Ecol Inform 75:102097

Muštra V, Rodríguez-Pose A (2021) Challenges and policies for middle-income trapped regions and cities. Reg Sci Policy Pract 13(3):438–440

Oktay D (2024) Sustainable urbanism and identity: a holistic perspective for future cities. Perspect Arch Urban 1(2):100016

Qorraj G, Hajrullahu A, Qehaja D (2024) Absorption capacity and the European integration of the Western Balkans. Reg Sci Policy Pract 100043.

Stanković J, Džunić M, Džunić Ž, Marinković S (2021) A multi-criteria evaluation of the European cities' smart performance: Economic, social and environmental aspects. Zbornik Radova Ekonomskog Fakulteta u Rijeci: Časopis Za Ekonomsku Teoriju i Praksu 35(2):519–550

Tošić D, Krunić N (2005) Urban agglomerations in the function of regional integration of Serbia in South-Eastern Europe. Bull Serb Geogr Soc 85(1):137–148

UN-Habitat (2022) World cities report 2022: envisaging the future of cities. https://unhabitat.org/sites/default/files/2022/06/wcr_2022.pdf

Vasilevska L, Zivkovic J, Vasilevska M, Lalovic K (2020) Revealing the relationship between city size and spatial transformation of large housing estates in post-socialist Serbia. J Housing Built Environ 35:1099–1121

Zhang Y (2021) China's green urbanization in the perspective of ecological civilization. Chin J Urban Environ Stud 9(01):2150001

Demographics

Chapter 2
Uncovering the Drivers of Urbanization in Serbia: An Econometric Investigation

Jelena J. Stanković, Marija Džunić, Srđan Marinković, and Ivana Marjanović

Abstract This study examines the factors influencing the attractiveness of 24 Serbian cities from 2011 to 2021, aiming to address gaps in understanding beyond economic determinants. Employing panel regression analysis on a sample of 24 Serbian cities from 2011 to 2021, we investigate the impact of employment, wages, transportation, culture, crime, healthcare, and education on urban attractiveness. Our findings imply the significance of employment opportunities, cultural amenities, and transportation accessibility in enhancing urban attractiveness. As expected, high crime rates significantly reduce city's appeal, emphasizing the crucial role of safety in urban development. Healthcare and education did not emerge as significant determinants of urban appeal in our analysis. This study's novelty lies in its comprehensive examination of diverse factors shaping urban attractiveness. By highlighting the multifaceted nature of urban attractiveness, our research offers insights for more holistic urban planning strategies, promoting safer, more vibrant, and economically resilient cities. The insights could be useful for policymakers and urban planners striving to create cities that are not only economically prosperous but also liveable and appealing to residents.

Keywords Urbanization · Urban attractiveness · Human capital · Panel data · Panel regression

2.1 Introduction

Urbanization serves as a pivotal catalyst for modernization and economic advancement, and it appears to be an inescapable trajectory in the progression of human society (Beall and Fox 2009). In the first half of the twentieth century, the majority of the global population resided in rural areas, with towns and semi-dense areas

J. J. Stanković (✉) · M. Džunić · S. Marinković · I. Marjanović
Faculty of Economics, University of Niš, Niš, Serbia
e-mail: jelena.stankovic@eknfak.ni.ac.rs

J. J. Stanković et al. (eds.), *Regional and Urban Challenges and Opportunities for Data-Policy Interactions*, The Urban Book Series,
https://doi.org/10.1007/978-3-032-08592-4_2

following closely behind. On the other hand, cities did not exhibit such a high population density. However, rapid growth in urban populations led to a more balanced distribution by 1965, with the global population evenly spread across rural areas, towns and semi-dense areas, and cities (UN-Habitat 2022). By the end of the twentieth century the order had reversed, with most people living in cities, followed by towns and semi-dense areas, and rural areas having the lowest population density. During that time, the labour force has undergone a swift transition away from traditional sectors and rural regions, gravitating towards employment opportunities in urban areas, causing an increase in the size of the cities. In 2020, nearly half of the global population resided in cities, while 29% lived in towns and semi-dense areas, and 22% lived in rural areas (UN-Habitat 2022). This shift signifies the increasing trend of urbanization and the concentration of populations in urban centres over the years. The expansion of urban population is, to a certain extent, shaped by the arrival of individuals aiming to alleviate poverty, escape from conflict, or mitigate economic and environmental risks (Šantić et al. 2023). Furthermore, a process of urbanization is projected to continue over the next three decades, with the urban population expected to increase from 56 per cent in 2021 to 68 per cent by 2050 (UN-Habitat 2022). Moreover, 96% of future urban growth is anticipated to occur in developing countries, offering them opportunities for sustainable economic development (Guo et al. 2022). This is particularly pertinent to the Western Balkans, where, according to United Nations projections, the proportion of the population residing in urban areas is anticipated to escalate by 2050, surpassing the global average (Gursoy and Kodaz 2021).

Urbanization is a complex process involving the migration of people from rural to urban areas in search of better opportunities and amenities. It leads to significant demographic and spatial changes, with cities becoming hubs for economic exchange and knowledge dissemination. However, urbanization also brings challenges such as resource depletion, environmental degradation, and uneven spatial expansion. Increased energy demand contributes to carbon emissions and climate change, while dense populations make urban areas more vulnerable to natural disasters and disease outbreaks. Effective urban planning is crucial to address these challenges and ensure sustainable development in urban areas.

Further research is imperative to explore the primary drivers of urbanization, offering a holistic understanding essential for creating effective policies and sustainable strategies for shaping the evolving urban landscape. Understanding these drivers facilitates proactive measures to optimize the benefits of urbanization and minimize its adverse effects. Well-planned urbanization not only spurs economic growth and innovation but also addresses social disparities and promotes environmental sustainability. Particularly in developing nations, urbanization presents avenues to alleviate poverty, inequality, and infrastructural deficiencies, offering improved access to essential services and fostering economic prosperity.

To support a more comprehensive analysis of the urbanization process in the Republic of Serbia, multidimensional statistical data from 24 Serbian cities spanning from 2011 to 2021 were obtained. By utilizing panel data and focusing on several indicators, panel regression is performed to examine the relationship between these factors and urbanization. This approach allows for a more in-depth and nuanced understanding of the dynamics of urbanization in the Republic of Serbia.

The chapter provides scientific contribution in two key aspects. Firstly, it contributes to the theoretical understanding of urbanization by providing a comprehensive background of urban attractiveness and explaining the factors that drive this process. Secondly, it contributes to the existing body of literature on urbanization by conducting empirical research using data from cities in the Republic of Serbia.

The chapter will be organized as follows: starting with introductory remarks, an examination of urbanization drives and challenges followed by patterns and trends specific to the Republic of Serbia will be provided. In the third section, a concise review of relevant scholarly literature on urban attractiveness will be offered. Subsequently, the next section will elaborate on the development and methodology of the proposed model, including a discussion on the data sources utilized in this study and the presentation and analysis of primary results derived from the conducted econometric analysis. Finally, the paper will offer concluding remarks.

2.2 Urbanization—Drivers and Challenges

Urbanization represents a multifaceted phenomenon involving the establishment and evolution of urban areas that possess distinct urban attributes (Gu 2019). Concurrently, urbanization stimulates the migration of individuals from rural communities towards cities, leading to the relinquishment of agricultural pursuits in favour of alternative endeavours (Busso et al. 2021). Specifically, urbanization is the process by which large numbers of people migrate from rural areas to urban settings, resulting in significant demographic and spatial changes. This phenomenon involves the movement of individuals and families seeking better economic opportunities, improved living conditions, and access to various amenities and services that are typically available in urban areas. Haisch and Müller (2015) state that urban regions serve as hubs for the exchange of economic commodities, human mobility, and knowledge dissemination and that they embody the focal points of progress and catalysts for transformative economic and social shifts.

Urbanization is characterized by a rise in the concentration of population within cities, concurrent with the expansion of urban land and infrastructure, as well as an elevation in per capita income (Guo et al. 2022). Multidimensionality of urbanization is reflected in the fact that the process of urbanization encompasses both the augmentation of population concentration in cities propelled by factors such as natural increase and in-migration. Additionally, it involves the examination of spatial morphological changes that define the patterns of such population growth, as emphasized by Hölscher and Frantzeskaki (2021). This population influx necessitates the development of housing, infrastructure, transportation systems, and other essential facilities to accommodate the growing urban population. Consequently, urban settings undergo physical changes, including the construction of new buildings, roads, parks, and other urban infrastructure. Nevertheless, the rapid process of urbanization has led to various challenges encompassing the depletion of resources, environmental degradation, uneven spatial expansion of cities, and a significant urban–rural disparity (Zhang 2021; Chen et al. 2023). The increased demand for energy and transportation in urban settings contributes to carbon emissions and climate change (Grimmond 2007). Furthermore, urban areas may also be more susceptible to natural disasters, such as floods or earthquakes, due to their dense populations and infrastructure vulnerabilities (Albrito 2012). In addition, the proximity of individuals in urban settings can facilitate the spread of communicable diseases (Kuddus et al. 2020).

Acknowledging these challenges, urban planning and governance play crucial roles in addressing them. There is an obvious need for further research to identify the key factors that drive urbanization. It is crucial to gain a comprehensive understanding of the drivers of urbanization from various perspectives to comprehend the challenges and opportunities they present. This knowledge will enable urban planners and local authorities to develop more efficient and sustainable policies to guide the process of urbanization. Understanding the factors that drive urbanization is of utmost practical importance for effective urban management and the formulation of policies on a global scale. By comprehending these drivers, local authorities can make informed decisions, implement appropriate strategies, and address the challenges and opportunities associated with urbanization. Through gaining a deeper understanding of the drivers of urbanization, policymakers can develop comprehensive and tailored policies that promote sustainable urban development, address social and economic disparities, and improve the quality of life for urban residents. This knowledge allows for the identification of potential risks and opportunities, enabling proactive measures to be taken to mitigate negative impacts and maximize the benefits of urbanization. Furthermore, understanding the drivers of urbanization provides valuable insights into the complex interplay between economic, social, and environmental factors. It helps in identifying the underlying causes of urban growth, such as population dynamics, economic forces, technological advancements, and social aspirations. This knowledge is essential for designing effective urban planning strategies, infrastructure development, and resource allocation to meet the evolving needs of urban populations. By adopting strategic approaches to urbanization, cities can become hubs for economic growth, innovation, and social progress. Well-planned urbanization ensures the efficient use of resources, promotes equitable access to services and opportunities, and fosters environmental sustainability (Chen et al. 2023). In developing countries, urbanization offers opportunities to address poverty, inequality, and inadequate infrastructure (Cobbinah et al. 2015). It can enable the provision of essential services such as education, healthcare, and clean water to larger populations, as well as create employment opportunities and drive economic growth.

2.3 Urbanization Patterns and Trends in the Republic of Serbia

Cities in the Central and Eastern European region have encountered significant challenges in adapting to the new economic and institutional landscape following the transition from a centrally planned to a market economy. This transformation has resulted in various urban issues, including the collapse of large state-owned industrial enterprises, high unemployment rates, poverty, deterioration of residential areas, infrastructural decay, and unused urban land. Following the regime shift, societies are characterized not only by very low levels of generalized trust and trust in institutions but also by a poor level of civil society development (Golubović et al. 2014). While there are certain unique aspects to the Republic of Serbia's development, its cities have experienced similar post-socialist urban transformations as those observed in other countries in Central and Eastern Europe (Vasilevska et al. 2020). The political and economic turmoil experienced by Serbia in the 1990s, coupled with the transition from a centrally controlled and urban-centric industrialization model to a market economy, has engendered substantial transformations in the evolution of cities. These urban areas have grappled with deteriorating demographic and economic vigour as a consequence. In the context of the Republic of Serbia, the phenomenon of urban sprawl has resulted in several problematic aspects. These encompass regional imbalances, polarization, the need for the revitalization of residential areas, and the imperative to preserve green spaces (Maksin-Mićić 2008).

The urban settlement network in Serbia exhibited significant disparities in terms of spatial, demographic, functional, and hierarchical development (Đurkin et al. 2021). The settlement framework of the Republic of Serbia encompasses a comprehensive assemblage of 6158 settlements (Statistical Office of the Republic of Serbia 2023). This settlement network exhibits a heterogeneous and disproportionate nature, characterized by an uneven distribution of settlements, population, centre accessibility, highly skilled labour, and the proportion of employed individuals (Srnić et al. 2021). Furthermore, small-scale settlements dominate the size structure of the overall settlement network with urban settlements comprising just above 3% of the settlement network (Statistical Office of the Republic of Serbia 2023). Nevertheless, there is a pronounced concentration of both population and activities in urban centres, with approximately 60% of the population residing in urban settlements (Srnić et al. 2021). Uneven concentration of population and varying degrees of economic development among different categories of cities hindered polycentric and balanced regional urban growth (Tošić and Krunić 2005). Within the network of urban centres, there is an evident disparity with merely 28 centres in Serbia having a population exceeding 30,000, while only four centres, namely Beograd, Novi Sad, Niš, and Kragujevac surpass the population threshold of 100,000. The spatial manifestation of the urbanization process revealed a development gap between cities of different ranks within the Serbian urban hierarchy (Đurkin et al. 2021). This indicates that the process of urbanization was not directed in a timely manner towards achieving a more balanced spatial distribution. Consequently, the phenomenon of demographic polarization in urban areas has led to the emergence of two distinct zones: the first characterized by a concentration of population in areas adjacent to major development axes, and the second marked by urban depopulation (Spasovski et al. 2012). According to Krunić et al. (2021), among the various difficulties that urban centres in Serbia encounter, two challenges are notably prominent: depopulation and the declining functional capabilities within these urban areas. Uneven demographic development and transitional economic transformation have resulted in varying levels of attractiveness among urban areas in the Republic of Serbia (Đurkin et al. 2021). Confronted with depopulation, the majority of cities in Serbia are compelled to formulate development policies focused on enhancing their attractiveness. This endeavour can only be achieved through thorough investigation of the determinants contributing to the appeal of cities.

2.4 Understanding the Concept of Urban Attractiveness

Cities are widely recognized for their multifaceted roles in societies across the globe (Scott and Storper 2015). They serve as centres of technological innovation and economic growth, driving progress and development in many nations (Stanković et al. 2021a, b). Cities act as hubs for innovation, attracting skilled professionals, entrepreneurs, and investors (Labrianidis et al. 2022). They offer abundant opportunities for employment, education, and cultural enrichment (Stanković et al. 2022). The city is an intricate and multifaceted entity designed for its inhabitants (Kourtit et al. 2021). Attracting highly skilled workers is pivotal for urban perspectives as the economic advancement of cities is dependent upon the expertise of such individuals (Buch et al. 2017; Stanković et al. 2021a, b). The attractiveness of cities is commonly linked to their economic successes, yet there is a growing consensus that other factors warrant consideration as pertinent (Marchesani 2023). Urban attractiveness is a measure of a city's capacity to fulfil conditions that discourage current residents from relocating elsewhere and encourage prospective residents to consider moving there (Fertner et al. 2015). One of the ways to evaluate relative attractiveness of a specific location can be by assessing the rise in the number of inhabitants (Partridge and Rickman 2003). Examining urban growth and comprehending its fundamental determinants is crucial for anticipating its far-reaching impacts (Marshall 2007). Population movements are influenced by a broad spectrum of pull factors extending beyond just employment opportunities or recreational amenities. These factors include intangible aspects such as social and cultural vitality, safety, inclusivity, and the availability of high-quality public services (Servillo et al. 2012). Several studies have highlighted the significance of spatial amenities and infrastructure for attracting highly educated and active individuals, commonly referred to as the "creative class" (Florida 2002). Nevertheless, beyond the realm of tangible structures and facilities, the functioning of modern cities is progressively influenced by the existence and calibre of social infrastructure (Caragliu and Del Bo 2012). The European Commission (2006) emphasizes key aspects of urban attractiveness, including a robust economic structure, accessibility to public services and institutions, efficient mobility and transportation, vibrant technological, cultural, and tourist environments, and an appealing natural and physical setting.

In today's evolving landscape, characterized by increased commuting, lack of affordable housing, and growing concerns about sustainable development, stakeholders are increasingly turning to measures such as quality of life, well-being, and sustainable living as key indicators of cities' attractiveness (Kwiatek-Sołtys and Mainet 2014). On the other hand, cities can also be marked by stark social and economic disparities. Rapid urbanization can lead to the concentration of poverty and the marginalization of certain communities. Inadequate housing, limited access to basic services, and unemployment can exacerbate social inequalities within urban areas. Hence, it is essential to thoroughly evaluate the factors of urban attractiveness and incorporate them into urban development policies to mitigate the adverse effects of urbanization.

Disparities in available amenities account for variations in urban areas' attractiveness (Buch et al. 2014). The attractiveness of a place is often assessed by the extent to which its features meet the preferences, needs, and desires of its residents (Barreira et al. 2019). Conversely, if those features do not meet the expectations, a place's attractiveness diminishes. According to Miot (2015), efforts by public authorities to enhance the residential functions of a location can greatly contribute to its attractiveness, fostering both the attachment and appeal of its inhabitants. The attractiveness of a city is closely linked to its ability to draw in or retain key demographic groups from less appealing areas, including residents, entrepreneurs, investors, and tourists. Each demographic group's retention or attraction adds value to the area in various ways (Bruneckienė et al. 2016): drawing in residents boosts consumption, enhances demographic diversity, and fosters community cohesion; attracting investors stimulates business growth and economic prosperity, lowers unemployment rates, and raises the area's profile; enticing tourists generates revenue in the tourism sector, thereby reinforcing local financial health.

Locational attributes serve as either push or pull factors, deterring or enticing potential population growth. These regional characteristics encompass labour market conditions, including employment rates and wage levels, as well as social and cultural amenities and public services. The attributes of a city associated with each of these domains may be perceived as either advantageous or disadvantageous in the competitive arena among urban regions (Stanković et al. 2017). For the assessment of urban attractiveness, in this chapter several factors are considered: employment opportunities, wage levels, quality of healthcare, safety issues, quality of infrastructure, quality of education and cultural amenities. All those domains are not only important motives for making decision about rural–urban migrations but also determine a complex of substitute and complement relations between cities themselves.

The attractiveness of a location often correlates with economic considerations, particularly the availability of services and employment opportunities (Barreira et al. 2019). Sohaib et al. (2022) confirm that employment opportunities represent an important factor that influences the attractiveness of cities. Aguayo and Guisan

(2004) demonstrate a causal link between employment opportunities and population migration. Regions with stronger economic activity tend to create more jobs, thereby drawing in skilled workers (Beckers and Boschman 2019). Therefore, it is expected that higher employment rate levels increase urban attractiveness.

Furthermore, wage level stands as an important factor influencing both internal and international migration, directly impacting the population's well-being. It not only prompts working-age individuals to relocate but also influences their family members' decisions to move to different cities (Bruneckienė et al. 2016). Butcher and Ebertz (2009a) assert that the minimum wage workers can earn in the new city has a significant influence on their decision when moving. However, the causality may be exactly the opposite one, with migrations influencing wage and income differences, what adds to the complexity of the relationship (Yin and Su 2022). Overall, we expect that the level of average wage should exert a positive influence on urban attractiveness.

When it comes to social amenities, the availability of healthcare is considered as important factor of urban attractiveness. Namely, the influence of the urban environment on residents' health can be twofold. On the one hand, social and health services are frequently more accessible in urban areas than in non-urban regions (Vlahov and Galea 2002), potentially enhancing the health and well-being of urban residents. The accessibility of medical professionals may serve as an indicator of the quality of healthcare within a specific city (Webster and Sanderson 2013). In this chapter the number of inhabitants per one physician will be used as proxy of the quality of healthcare. The lower level of this indicator would imply higher quality of healthcare which should lead to increased urban attractiveness.

The prevalence of crime in a city is a fundamental indicator that reflects the overall safety of living conditions. Cities that are perceived as safe attract businesses, investments, and tourism, consequently fostering job creation and economic growth (Moore and Woodcraft 2019). Elevated crime rates can discourage economic activities, diminish property values, and impede efforts aimed at urban revitalization (Rupp et al. 2020). Furthermore, high crime rates are anticipated to diminish the attractiveness and desirability of a city for its residents, consequently impacting its net migration rate (Buch et al. 2014). The crime rate frequently influences individuals' decisions to relocate (Buettner and Ebertz 2009b). In this chapter, the crime rate is operationalized as the aggregate number of legally convicted adults and minors, and it is hypothesized to exert a negative influence on urban attractiveness.

Accessibility pertains to individuals' capacity to access a wide range of geographically dispersed attractions, activities, and amenities (Solá et al. 2018). It is widely recognized that disparity in proximity to various amenities between urban, sub-urban, and rural areas impacts neighbourhood satisfaction, as reflected in housing prices (Shin et al. 2024; Soltani et al. 2024). However, the geographical proximity affects accessibility not directly but via the transportation infrastructure. The improvement of transportation infrastructure facilitates the accessibility of rural areas to cities, potentially prompting the migration of productive capital and skilled labour from rural regions to urban centres over time (Banerjee et al. 2020). Garcia-Lopez (2019) conducted a study on the influence of highways on residential sprawl and found that the presence of highways correlates with an expansion of residential land area. Similarly, Li et al. (2023) found an existence of rapid transit positively connected only to the attractiveness of sub-urban districts, opposite to the inner-city area, and explained the difference by the disparities of spatial quality, economic development, and public facilities and amenities. Thus, finding a right proxy for transportation infrastructure is a challenging task. Being constrained with data availability, this chapter represents quality of infrastructure by the length of roads within a city. Therefore, the length of roads should have a positive influence on urban attractiveness.

The provision of cultural and social infrastructure and services, including educational institutions, stands as one of the primary factors contributing to the attractiveness of a city. Conversely, the absence of such amenities diminishes the city's allure and erodes its image. Location factors play a crucial role in persuading residents to remain in a city and in attracting newcomers (Glaeser et al. 2014). Investing in the quality of primary and secondary education, which serves as a prerequisite for higher education, can also be viewed as an effective strategy to mitigate human capital loss (Nelle 2016). Glaeser et al. (2004) discovered that offering fundamental quality education can both generate and attract educated individuals. On the other hand, Lehtonen (2021) examined the effects of primary school closures on population numbers in communities in Finland. The findings suggest that following school closures, the population of the surrounding community experienced a greater decrease compared to the period before the closures. In this chapter, the quality of education is proxied by the number of secondary schools. Therefore, the number of secondary schools within a city should have a positive influence on population growth.

Availability of cultural content contributes to an improved quality of life for urban residents by fostering a pleasant living environment and alleviating the adverse impacts of extensive urbanization (Wills-Herrera et al. 2009; Bandarin and Van Oers 2012). Cultural amenities have been strategically highlighted as pivotal components and leveraged as instruments to appeal to the creative class (Hichri et al. 2023). The clustering of cultural amenities and artistic communities is viewed as a significant indicator of urban tolerance, a factor particularly favoured by the mobile creative class (Florida 2002). In this chapter, the availability of cultural amenities is represented by the number of cinemas within a city. It is anticipated that cultural amenities will exert a positive influence.

2.5 Econometric Model and Discussion

To evaluate the determinants of urban attractiveness, multiple regression analysis on panel data was applied. Panel data represent a fusion of time series data and cross-sectional data, offering both temporal and cross-sectional dimensions simultaneously (Biørn 2017). The dependent variable is population growth represented as the difference between the population in the year t and $t-1$ and serves as proxy for urban attractiveness. All regressors are treated as predetermined to address the potential endogeneity of the explanatory variables.

The regression model is formed as follows:

$$\begin{aligned}\text{Growth}_{i,t} = {} & \beta_0 + \beta_1\text{Wage}_{i,t-1} + \beta_2\text{Employment}_{i,t-1} + \beta_3\text{Health}_{i,t-1} \\ & + \beta_4\text{Transport}_{i,t-1} + \beta_5\text{Culture}_{i,t-1} + \beta_6\text{Education}_{i,t-1} \\ & + \beta_7\text{Crime}_{i,t-1} + \varepsilon_{i,t}\end{aligned}$$

where:

$\beta_0, \beta_1, \ldots, \beta_7$ represent intercept and slope coefficients,

$\varepsilon_{i,t}$ disturbance term,

$i = 1, \ldots, 24$,

$t = 2, \ldots, 11$.

The description of variables and expected sign is provided in Table 2.1.

The panel data regression model considers both cross-sectional and temporal dimensions, which prompts the need to account for the interdependence of selected variables over time and across different entities. Establishing an appropriate model specification involves assessing the variability of regression parameters, to detect and evaluate individual and temporal effects. The outcomes of the Breusch-Pagan

Table 2.1 Definition and expected sign of explanatory variables

Variable	Definition	Source	Expected sign
Wage	Average wage (tax and contributions excluded) per employee in local currency. A monthly average for a reporting year	Statistical Office of the Republic of Serbia	+
Employment	Registered employment by municipality of residence	Statistical Office of the Republic of Serbia	+
Health	Number of inhabitants per 1 physician	Statistical Office of the Republic of Serbia	–
Transport	The length of roads (paved road surface) in km	Statistical Office of the Republic of Serbia	+
Culture	Number of cinemas in the city	Statistical Office of the Republic of Serbia	+
Education	Number of colleges and faculties in the city	Statistical Office of the Republic of Serbia	+
Crime	Legally convicted adults and minors per 1000 inhabitants	Statistical Office of the Republic of Serbia	–

Lagrange multiplier test and the joint F-test refute the pooled OLS assumption (Table 2.2). Conversely, the results of the Hausman test support the model with fixed effects (Table 2.2). Furthermore, it is necessary to test assumptions related to homoscedasticity, multicollinearity, and cross-sectional dependence. The findings suggest no sign of multicollinearity (Mean VIF = 4.68); however, the model exhibits heteroscedasticity and cross-sectional dependence (Table 2.2). To address this issue, Driscoll-Kraay standard errors are utilized.

The results of the estimated model are presented in Table 2.3.

Table 2.2 Model specification and diagnostics test results

Test	Value	Significance
Breusch-Pagan Lagrange multiplier test	615.48	0.0000
F-test	112.43	0.0000
Hausman's test	36.59	0.0000
Modified Wald's test for groupwise heteroscedasticity in fixed effect regression model	8,940.66	0.0000
Pesaran's test of cross-sectional independence	10.400	0.0000

Table 2.3 Fixed panel regression results

Explanatory variables	Estimates
Employment	75.63***
	(8.008)
Wage	0.107**
	(0.0425)
Health	−5.644
	(8.396)
Transport	6.827*
	(3.558)
Culture	0.0127***
	(0.00183)
Crime	−3.528***
	(0.982)
Education	144.3
	(235.4)
Constant	19,622***
	(4,008)
Observations	240
Number of cities	24
R-squared	0.426

Standard errors in parentheses, *** $p < 0.01$, ** $p < 0.05$, * $p < 0.1$

The results indicate that Employment, Wage, Transport, Culture and Crime represent significant determinants of the attractiveness of Serbian cities, where variables Employment, Wage, Transport, Culture have positive influence, while the influence of Crime variable is negative. These findings are in line with the assumptions. Namely, a larger city offers residents a better quality of life by increasing income and job opportunities. Several research highlight the significance of job opportunities in shaping the appeal of urban areas. Additionally, the cultural offerings of cities play a crucial role in their development. An abundance of cultural content not only enhances the quality of life for city dwellers but also mitigates the negative effects of rapid urbanization, creating a more pleasant living environment. Regarding Transportation variable, accessibility represents an important factor of cities attractiveness as it refers to enabling individuals to reach various attractions, activities, and amenities spread across different locations. Having access to people, services, and employment opportunities constitutes a fundamental element in the economic dynamism of a city and its residents' quality of life. Concerning the Crime variable, the safety of urban areas holds significant importance for their residents. Safe cities foster an environment conducive to social interactions and community engagement. Safety plays a pivotal role in driving economic development and prosperity within urban settings. Urban areas with a reputation for safety attract businesses, investments, and tourists, thereby stimulating job opportunities and economic development. On the contrary, high levels of crime can deter economic activity and decrease perceived quality of life in urban areas, leading to the decline of urban attractiveness. There was no compelling statistical evidence of importance of the Health and Education proxy used in this study.

Considering the above results, specific recommendations can be made for policymakers to ensure balanced regional development. Bearing in mind that inadequate development policies have led to regional imbalances, further polarization may lead to their aggravation. To mitigate this and promote balanced regional growth, polycentric development is proposed as a necessary planning measure (Živanović et al. 2019). Increasing the attractiveness of cities, particularly those with negative migration balances, can help balance regional development. In that sense, adequate development policies should be created based on analyses of urban attractiveness. Investments, along with state incentives, could encourage population influx into medium-sized and small cities, enhancing their attractiveness and alleviating existing regional disparities (Živanović et al. 2019). Decentralized planning, involving the establishment of small and medium-sized enterprises in small towns linked to industries in larger cities, is also recommended (Živanović et al. 2021). Encouraging entrepreneurship through training can further promote job creation. Additionally, cities can invest in development programmes aimed at aligning skills with local industry needs (Li 2022). Furthermore, to improve urban accessibility and connectivity, investments in public transport networks, as well as improvements to pedestrian and cycling infrastructure, in order to promote active transport and reduce car dependence (Zakharov and Fadyushin 2021). Enhancing a city's image can be also achieved by supporting and investing in the arts and culture sectors, as the creative class is attracted to cultural offerings (Cunningham et al. 2021). Promoting cultural events, festivals, and partnering with cultural organizations can enhance a city's cultural content. Ensuring the safety of residents is also crucial, which can be achieved through initiatives aimed at improving public safety and building trust between law enforcement agencies and residents, such as investing in crime prevention programmes (O'Malley and Smith 2022).

2.6 Concluding Remarks

Cities represent dynamic hubs of innovation, economic growth, and cultural exchange, attracting skilled professionals and fostering diverse opportunities for employment, education, and enrichment. However, their attractiveness depends on a number of factors beyond mere economic prosperity, including access to quality healthcare, safety, transportation infrastructure, and cultural amenities. The concept of urban attractiveness implies that mobility of people is not solely dictated by the tangible factors present in each area. Instead, it highlights the possibility of distinct amenities where individual choices are influenced by a blend of spatial elements and social requirements, thereby adding complexity to the mobility equation (Servillo et al. 2012). Over time, comprehension of population mobility dynamics has evolved from a belief that economic factors predominantly dictate population movements to one that embraces a broader concept related to the quality of place. Therefore, cities should offer not only financial incentives but also other amenities such as a conducive living environment, rich cultural offerings, and robust infrastructure to attract and retain a skilled workforce. Nowadays, cities must adopt a proactive approach with the aim of positioning themselves as central hubs of human capital which can be achieved by fostering favourable conditions for the population.

Based on the analysis of panel data from 24 Serbian cities spanning the period from 2011 until 2021, the conclusions unequivocally demonstrate that urban attractiveness transcends the populace's perception of financial prosperity. Rather, it also pivots towards factors such as cultural amenities, safety, and accessibility. While salary remains a significant factor in urban attractiveness, residents equally prioritize urban accessibility, the availability of cultural content, and the safety of the cities themselves, reflected in lower crime rates.

The ability to attract a workforce is paramount for the future prosperity of urban centres. Our research outcomes yield significant insights for municipal authorities and urban planning professionals. The findings underscore the potential for adjusting multiple factors influencing individuals' decisions to relocate to or remain within a city. The analysis results offer valuable guidance to local authorities for identifying the factors contributing to urban attractiveness and for creating tailored urban development policies to capitalize on these opportunities. This highlights the flexibility local governments possess in formulating tailored strategies aimed at enhancing the city's attractiveness to prospective employees.

The chapter makes a dual contribution: theoretically, it enriches the existing literature on the concept of urban attractiveness, and empirically, it provides research on the determinants of urban attractiveness, illustrated through examples from cities in the Republic of Serbia. Nevertheless, the research encounters certain limitations, such as the unavailability of data for all cities in the Republic of Serbia and the omission of data concerning the environmental performance of cities due to their unavailability for the whole observed period. Future research efforts could focus on addressing these identified shortcomings, particularly by incorporating data related to the pollution and wastewater in the cities to obtain a broader picture of determinants of urban attractiveness.

Funding This study was supported by the European Union's Horizon Europe research and innovation programme WIDERA-2021-ACCESS-02 under grant number 101059994—UR-Data.

References

Aguayo E, Guisan MC (2004) Employment and population in European union: econometric models and causality tests. Docum Ser Econ Dev (80)

Albrito P (2012) Making cities resilient: increasing resilience to disasters at the local level. J Bus Contin Emer Plan 5(4):291–297

Almulhim AI, Cobbinah PB (2023) Can rapid urbanization be sustainable? The case of Saudi Arabian cities. Habitat Int 139:102884

Bandarin F, Van Oers R (2012) The historic urban landscape: managing heritage in an urban century. John Wiley & Sons

Banerjee A, Duflo E, Qian N (2020) On the road: access to transportation infrastructure and economic growth in China. J Dev Econ 145:102442

Barreira AP, Nunes LC, Guimarães MH, Panagopoulos T (2019) Satisfied but thinking about leaving: the reasons behind residential satisfaction and residential attractiveness in shrinking Portuguese cities. Int J Urban Sci 23(1):67–87

Beall J, Fox S (2009) Cities and development. Routledge

Beckers P, Boschman S (2019) Residential choices of foreign highly skilled workers in the Netherlands and the role of neighbourhood and urban regional characteristics. Urban Stud 56(4):760–777

Biørn E (2017) Econometrics of panel data: methods and applications. Oxford University Press

Bruneckienė J, Zykienė I, Stankevičius V (2016) Critical analysis of city attractiveness factors in Lithuania-Poland cross-border regions: the viewpoints of businessmen and youth. J Geogr Polit Soc 6(2):45–58

Buch T, Hamann S, Niebuhr A, Rossen A (2014) What makes cities attractive? The determinants of urban labour migration in Germany. Urban Stud 51(9):1960–1978

Buch T, Hamann S, Niebuhr A, Rossen A (2017) How to woo the smart ones? Evaluating the determinants that particularly attract highly qualified people to cities. J Urban Aff 39(6):764–782

Buettner T, Ebertz A (2009a) Spatial implications of minimum wages. Jahrbücher Für Natl Und Stat 229(2–3):292–312

Buettner T, Ebertz A (2009b) Quality of life in the regions: results for German Counties. Ann Reg Sci 43:89–112

Busso M, Chauvin JP, Herrera N (2021) Rural-urban migration at high urbanization levels. Reg Sci Urban Econ 91:103658

Caragliu A, Del Bo C (2012) Smartness and European urban performance: assessing the local impacts of smart urban attributes. Innov Europ J Soc Sci Res 25(2):97–113

Chen R, Chen Y, Lyulyov O, Pimonenko T (2023) Interplay of urbanization and ecological environment: coordinated development and drivers. Land 12(7):1459
Cobbinah PB, Erdiaw-Kwasie MO, Amoateng P (2015) Rethinking sustainable development within the framework of poverty and urbanisation in developing countries. Environ Dev 13:18–32
Cunningham S, McCutcheon M, Hearn G, Ryan MD (2021) 'Demand' for culture and 'allied' industries: policy insights from multi-site creative economy research. Int J Cult Policy 27(6):768–781
Đurkin D, Antić M, Budović A (2021) Demographic and economic aspects of urban shrinkage in Serbia-typology and regional differentiation. Glas Srp Geogr Drus 101(2):43–78
European Commission (2006) Cohesion policy and cities: the urban contribution to growth and jobs in the regions. In: Communication from the commission to the council and parliament COM(2006) 385 final of 13.7.2006. European Commission, Brussels. https://eur-lex.europa.eu/LexUriServ/LexUriServ.do?uri=COM:2006:0385:FIN:EN:PDF
Fertner C, Groth NB, Herslund L, Carstensen TA (2015) Small towns resisting urban decay through residential attractiveness. Findings from Denmark. Geogr Tidsskr-Dan J Geogr 115(2), 119–132
Florida R (2002) The economic geography of talent. Ann Assoc Am Geogr 92(4):743–755
Garcia-López MÀ (2019) All roads lead to Rome… and to sprawl? Evidence from European cities. Reg Sci Urban Econ 79:103467
Glaeser EL, Ponzetto GA, Tobio K (2014) Cities, skills and regional change. Reg Stud 48(1):7–43
Glaeser EL, Saiz A, Burtless G, Strange WC (2004) The rise of the skilled city [with comments]. In: Brookings-Wharton papers on urban affairs, pp 47–105
Golubović N, Džunić M, Marinković S (2014) A comparative analysis of social capital in the Western Balkan countries. Industrija 42(2):71–92
Grimmond SUE (2007) Urbanization and global environmental change: local effects of urban warming. Geogr J 173(1):83–88
Gu C (2019) Urbanization: Processes and driving forces. Sci China Earth Sci 62:1351–1360
Guo J, Yu Z, Ma Z, Xu D, Cao S (2022) What factors have driven urbanization in China? Environ Dev Sustain 24(5):6508–6526
Gursoy O, Kodaz M (2021) Urbanization and environmental problems in the Western Balkans and policy recommendations. In: Korkut H (ed) Studies in urbanization and local governance in Balkan countries. Dobra knjiga, Sarajevo, pp 15–38
Haisch T, Müller U (2015) Defining and measuring urban regions: a sensitivity analysis. Pap Reg Sci 94(1):219–226
Hichri SM, Jamoussi HBO, Keraani W (2023) Rethinking the dimensions of the creative city: beyond urban amenity. In: The creative class revisited: new analytical advances. pp 111–152
Hölscher K, Frantzeskaki N (2021) Perspectives on urban transformation research: transformations in, of, and by cities. Urban Transform 3:1–14
Kourtit K, Nijkamp P, Wahlström MH (2021) How to make cities the home of people–a 'soul and body' analysis of urban attractiveness. Land Use Policy 111:104734
Krunić N, Gajić A, Srnić D (2021) How system of settlements in Serbia will change? Geographic determinants of spatial-functional relationships of urban and rural areas. International Scientific Conference GEOBALCANICA 2021. Geobalcanica Society, Skopje, North Macedonia, pp 461–465
Kuddus MA, Tynan E, McBryde E (2020) Urbanization: a problem for the rich and the poor? Public Health Rev 41:1–4
Kwiatek-Sołtys A, Mainet H (2014) Quality of life and attractiveness of small towns: a comparison of France and Poland. Quaest Geogr 33(2):103–113
Labrianidis L, Sykas T, Sachini E, Karampekios N (2022) Highly educated skilled migrants are attracted to global cities: the case of Greek PhD holders. Popul Space Place 28(3):e2517
Lehtonen O (2021) Primary school closures and population development–is school vitality an investment in the attractiveness of the (rural) communities or not? J Rural Stud 82:138–147

Li Y, Lin Y, Wang J, Geertman S, Hooimeijer P (2023) The effects of jobs, amenities, and locations on housing submarkets in Xiamen City, China. J Housing Built Environ 38:1221–1239. https://doi.org/10.1007/s10901-022-09984-8

Li L (2022) Reskilling and upskilling the future-ready workforce for industry 4.0 and beyond. Inform Syst Front 1–16

Maksin-Mićić M (2008) Sprawl and sustainable urban development in Europe. Arhit I Urban 22–23:77–83

Marchesani F (2023) Urban attractiveness and the competitive edge of the smart city. In: The global smart city. Emerald Publishing Limited, pp 117–135

Marshall JD (2007) Urban land area and population growth: a new scaling relationship for metropolitan expansion. Urban Stud 44(10):1889–1904

Miot Y (2015) Residential attractiveness as a public policy goal for declining industrial cities: housing renewal strategies in Mulhouse, Roubaix and Saint-Etienne (France). Eur Plan Stud 23(1):104–125

Moore HL, Woodcraft S (2019) Understanding prosperity in East London: local meanings and "sticky" measures of the good life. City Soc 31(2):275–298

Nelle AB (2016) Tackling human capital loss in shrinking cities: urban development and secondary school improvement in Eastern Germany. Eur Plan Stud 24(5):865–883

O'Malley P, Smith GJ (2022) 'Smart' crime prevention? Digitization and racialized crime control in a Smart City. Theor Criminol 26(1):40–56

OECD/UN-HABITAT/UNOPS (2021) Global state of national urban policy 2021: achieving sustainable development goals and delivering climate action. OECD Publishing, Paris, https://doi.org/10.1787/96eee083-en

Partridge MD, Rickman DS (2003) The waxing and waning of regional economies: the chicken–egg question of jobs versus people. J Urban Econ 53(1):76–97

Rupp LA, Zimmerman MA, Sly KW, Reischl TM, Thulin EJ, Wyatt TA, Stock JJP (2020) Community-engaged neighborhood revitalization and empowerment: busy streets theory in action. Am J Community Psychol 65(1–2):90–106

Šantić D, Langović M, Đorđević DŽ (2023) Urban sustainability through the lens of migration-case study: city of Leskovac, Serbia. Econ Themes 61(1):105–119

Scott AJ, Storper M (2015) The nature of cities: the scope and limits of urban theory. Int J Urban Reg Res 39(1):1–15

Servillo L, Atkinson R, Russo AP (2012) Territorial attractiveness in EU urban and spatial policy: a critical review and future research agenda. Europ Urban Reg Stud 19(4):349–365

Shin J, Newman G, Park Y (2024) Urban versus rural disparities in amenity proximity and housing price: the case of integrated urban-rural city, Sejong, South Korea. J Housing Built Environ. Advance online publication. https://doi.org/10.1007/s10901-023-10098-y

Sohaib K, Driss E, Youness J (2022) An econometric analysis of the determinants of small's cities' attractiveness: evidence from Moroccan Case. In: 2022 8th International conference on optimization and applications (ICOA). IEEE, pp 1–6

Solá AG, Vilhelmson B, Larsson A (2018) Understanding sustainable accessibility in urban planning: themes of consensus, themes of tension. J Transp Geogr 70:1–10

Soltani A, Zali N, Aghajani H, Hashemzadeh F, Rahimi A, Heydari M (2024) The nexus between transportation infrastructure and housing prices in metropolitan regions. J Housing Built Environ. Advance online publication. https://doi.org/10.1007/s10901-023-10085-3

Spasovski M, Šantić D, Radovanović O (2012) Historical stages in transition of natural replacement of the Serbian population. Glas Srp Geogr Društva 92(2):23–60

Srnić D, Krunić N, Gajić A (2021) Urban areas of Serbia–a new framework for spatial development. International scientific conference GEOBALCANICA 2021. Geobalcanica Society, Skopje, North Macedonia, pp 477–482

Stanković J, Džunić M, Džunić Ž, Marinković S (2017) A multi-criteria evaluation of the European cities' smart performance: economic, social and environmental aspects. Zb Rad Ekon Fak U Rijeci Časopis Za Ekon Teor I Praksu 35(2):519–550

Stanković JJ, Popović Ž, Marjanović I (2021b) Assessing smartness and urban development of the European cities: an integrated approach of Entropy and VIKOR. Mult Criteria Decis Mak Inf Age 25:69–97

Stanković JJ, Marjanović I, Drezgić S (2021a) Urban magnetism in the global city framework: exploring the link between urban functions and population growth

Stanković JJ, Džunić M, Marjanović I (2022) Towards an inclusive Europe: ranking European Countries based on social sustainability indicators. In: International conference on decision support system technology. Springer International Publishing, Cham, pp 84–96

Statistical Office of the Republic of Serbia (2023) Statistical Yearbook of the Republic of Serbia. https://publikacije.stat.gov.rs/G2023/Pdf/G20232056.pdf

Swiney C (2020) The urbanization of international law and international relations: the rising soft power of cities in global governance. Mich J Int'l l 41:227

Tošić D, Krunić N (2005) Urban agglomerations in the function of regional integration of Serbia in South-Eastern Europe. Bull Serbian Geogr Soc 85(1):137–148

UN-Habitat (2022) World cities report 2022 envisaging the future of cities. https://unhabitat.org/sites/default/files/2022/06/wcr_2022.pdf

United Nations (2016) New urban agenda. https://habitat3.org/wp-content/uploads/NUA-English.pdf

Vasilevska L, Zivkovic J, Vasilevska M, Lalovic K (2020) Revealing the relationship between city size and spatial transformation of large housing estates in post-socialist Serbia. J Housing Built Environ 35:1099–1121

Vlahov D, Galea S (2002) Urbanization, urbanicity, and health. J Urban Health 79:S1–S12

Webster P, Sanderson D (2013) Healthy cities indicators—a suitable instrument to measure health? J Urban Health 90:52–61

Wills-Herrera E, Islam G, Hamilton M (2009) Subjective well-being in cities: a multidimensional concept of individual, social and cultural variables. Appl Res Qual Life 4:201–221

Yin X-C, Su C-W (2022) Have housing prices contributed to regional imbalances in urban-rural income gap in China? J Housing Built Environ 37:2139–2156. https://doi.org/10.1007/s10901-022-09945-1

Zakharov D, Fadyushin A (2021) The efficiency of some activities for the development of urban infrastructure for public transport, cyclists and pedestrians. Int J Trans Dev Integ 5(2):136–149

Zhang Y (2021) China's green urbanization in the perspective of ecological civilization. Chin J Urban Environ Stud 9(01):2150001

Živanović Z, Tošić B, Nikolić T, Gatarić D (2019) Urban system in Serbia—the factor in the planning of balanced regional development. Sustainability 11(15):4168

Živanović Z, Tošić B, Nikolić T, Samardžić I, Dabović T, Jeftić M (2021) Urban shrinkage in Serbia: the domination of economic over environmental causes. FEB-Fresenius Environ Bull 13437

Data-Policy Interaction

The future of cities should be envisioned in alignment with the principles of the New Urban Agenda and the 2030 Agenda for Sustainable Development, with particular emphasis on Sustainable Development Goal 11. Launched in 2016, the Urban Agenda for the EU is a voluntary partnership between EU member states, cities, and other stakeholders. It focuses on several priority themes, including housing, air quality, and sustainable land use. The agenda encourages collaboration and the exchange of best practices to address urban challenges. The New Urban Agenda embodies a collective vision for a more promising and sustainable future. When approached with detailed planning and effective management, urbanization can serve as a potent catalyst for sustainable development, benefiting both developing and developed countries (United Nations 2016). In addition, other global frameworks such as the Paris Agreement on Climate Change and the Sendai Framework for Disaster Risk Reduction are also relevant for promoting sustainable urbanization (Swiney 2020; Almulhim and Cobbinah 2023). In the Republic of Serbia national urban policy is defined through Sustainable Urban Development Strategy of the Republic of Serbia 2030 adopted in 2019. National Urban Development Policy represents “a coherent set of decisions, guided by the national government through the process of co-operation of various actors in formulating a common vision and common goals, which are used to direct long-term transformative, productive, inclusive and resilient sustainable urban development” (OECD/UN-HABITAT/UNOPS 2021, p. 35).

Jelena J. Stanković is a Full Professor at the Faculty of Economics, University of Niš, Serbia in the scientific field of mathematics and statistics in economics. Her research focuses on quantitative economic analysis, especially in multi-criteria decision making for local economic development issues, urban labour market, and smart and sustainable development of urban areas. She is an active member if the Committee for Economic Sciences of the Serbian Academy of Science and Arts. She has been involved in a number of national and international projects as researcher and principal investigator. Prof. Stanković is an editorial board member of the journal ‘Review: Journal of Social Issues’, published by the University of Sarajevo, the ‘Journal of Economics and Business’ published by the University of Rijeka, and the ‘International Journal of Sustainable Agricultural Management and Informatics’ published in Switzerland. She has published numerous academic articles in international journals.

Marija Džunić is a Full Professor of Economic Theory and Political Economy at the Faculty of Economics, University of Niš, Serbia. Her areas of expertise are social determinants that shape economic outcomes with a focus on urban employment dynamics, social capital and democratic consolidation. She has extensive experience in teaching, research and project implementation. Currently, she is leading in EU-funded projects dissemination and outreach activities. Further, Prof. Džunić is a member of the presidency of the the Scientific Society of Economists in Serbia.

Srđan Marinković is a Full Professor of Banking and Fiance at the Faculty of Economics, University of Niš, Serbia. He holds a PhD from the University of Niš and has successfully finalised several doctoral studies courses at Aarhus School of Business and the University of Oslo. In addition, he conducted his postdoctoral research at the London School of Economics and Political Science. Furthermore, Prof. Marinković was a visiting fellow at a number of international institutions including Norwegian School of Economics, London School of Economics and Political Science and the University of Ljubljana. He is a member of the presidency of the National Scientific Society of Economists and is a former editor-in-chief of the academic journal ‘Economic Themes’. He has published widely well-established internationally.

Ivana Marjanović is a Teaching at the Faculty of Economics, University of Niš, Serbia. She holds a PhD in Economics and her research focuses on the application of quantitative methods with a focus on decision theory models and operational research in the field of economy. In addition, Dr. Marjanović has expierence in the application of mixed methods in urban studies for sustainable development. She has been involved in international research and Erasmus+ projects. She has published over 50 scientific papers.

Chapter 3
Exploring the Relationship Between Economic and Demographic Performance of Local Governments: An Analysis of Factors and Implications

Marina Čolig, Kristina Kregar, and Saša Drezgić

Abstract This paper aims to investigate the relationship between economic and demographic performance of local governments, using the Tiebout theorem as a framework. The theorem posits that individuals will relocate to jurisdictions that provide public goods and services that align with their preferences. Therefore, local governments that offer a diverse range of public goods and services can attract a wider range of residents, promoting economic growth. Demographic factors, such as age, income, and education levels, are also likely to influence economic and demographic performance. We argue that in less developed and emerging economies, residents prioritize job creation and appropriate wages over other factors. Thus, local governments in these economies should prioritize job creation to attract and retain residents and stimulate economic growth. Our study offers valuable insights for policymakers seeking to enhance economic and demographic performance in their local governments.

Keywords Tiebout theorem · Economic performance · Demographic factors · Job creation · Less developed economies

---

S. Drezgić (✉)
Faculty of Economics, University of Rijeka, Rijeka, Croatia
e-mail: sasa.drezgic@efri.uniri.hr

M. Čolig
Municipality of Zagreb, Croatia

K. Kregar · S. Drezgić
University of Rijeka, Rijeka, Croatia

K. Kregar
University of Trieste, Trieste, Italy

J. J. Stanković et al. (eds.), *Regional and Urban Challenges and Opportunities for Data-Policy Interactions*, The Urban Book Series,
https://doi.org/10.1007/978-3-032-08592-4_3

3.1 Introduction

The Tiebout theorem has long been a useful framework for understanding how the provision of public goods and services by local governments can affect economic growth and demographic composition. According to the theorem, individuals choose to relocate to areas that offer public goods and services that match their preferences. This relocation can lead to a diversity of residents that promotes economic growth in these areas. Of course, when discussing the theorem, one must keep in mind its rigid theoretical assumptions, which, among other things, do not take into account "constraints due to employment opportunities" and assume "that all persons live on dividend income" (Tiebout 1956). The theorem thus assumes that economic or market conditions do not affect consumers' and voters' jurisdictional choices through their preferences for public goods. In this paper, we examine the relationship between economic and demographic performance of local governments using the Tiebout theorem as a framework.

We focus on the interplay between demographics, economics and the efficiency of local governments in providing an optimal quantity and quality of public goods and services that lead to positive outcomes in terms of economic and population growth and overall welfare levels. The interplay between these factors and the provision of public goods and services is complex and requires thorough investigation to understand the relationship between economic and demographic performance. In this respect, our paper contributes to the existing literature by providing new theoretical insights into the relationship between economic and demographic performance.

In many less developed and emerging economies, job creation and decent wages for residents are top priorities. The reason for this is the high levels of unemployment and poverty in these regions. Local governments in these economies must therefore prioritize job creation to attract and retain residents and promote economic growth. But other factors, such as access to education and healthcare, can also be important in attracting residents and promoting economic growth. Demographic factors such as age, income, and education levels are also likely to influence economic and demographic performance. Therefore, it is important to examine the relationship between the provision of public goods and services and economic and demographic performance in these regions.

The main objective of this paper is to examine the relationship between the provision of public goods and services and economic and demographic performance in communities, assuming significant economic and demographic inequalities within the national area. We will analyze the validity of the Tiebout theorem in these regions and examine how demographic factors such as age, income, and educational attainment influence the relationship between economic and demographic performance. Our study offers valuable insights for policy makers seeking to improve the economic and demographic performance of their communities.

To summarize, this paper aims to contribute to the existing literature by examining the relationship between the provision of public goods and services and economic and demographic performance in local governments, using the Tiebout theorem as a framework. We argue that local governments in less developed and emerging economies should prioritize job creation to attract and retain residents and promote economic growth. We hope that our findings will provide valuable insights for policy makers seeking to improve the economic and demographic performance of their local authorities.

The next chapter looks at the provision of local public goods and services and economic performance, followed by the third chapter, which looks at demographic performance. The policy implications represent an integration of the two chapters and provide suggestions for local government policy. The conclusions offer some potential research opportunities.

3.2 Tiebout Model and Economic Performance

The Tiebout theorem, first proposed by Charles Tiebout in 1956, states that individuals move to areas that offer public goods and services that match their preferences (Tiebout 1956). This relocation can lead to a diversity of residents that promotes economic growth in these areas. The theorem is a useful framework for understanding how the provision of public goods and services can influence economic growth. Tiebout bases his theorem on the research of Musgrave (1939) and Samuelson (1954, 1955), who argued that there is no appropriate market solution that determines the level of consumption (cost) of public goods at the federal level, from which it follows that a significant portion of the realized social product is allocated in a non-optimal way compared to the allocation of private goods. However, Tiebout (1956) pointed out that this is not the case for the cost of public goods at the local level. Tiebout argues that the fundamental problem identified by Musgrave and Samuelson, that there is no mechanism to force the consumer (voter) to express his true preferences, does not work at the local level. Musgra1ve and Samuelson's conclusion arises from the fact that the rational consumer does not want to reveal his preferences in order to benefit from the public goods on offer and at the same time avoid paying taxes. However, Tiebout (1956) stated that consumers (voters) will move to the municipality that best suits their preferences for public goods and services. A larger number of municipalities means a greater difference between them, but also provides a greater opportunity for consumers to fully satisfy their preferences for public goods and services. This mechanism of choosing the most appropriate local community is known as "voting with one's feet."

Boadway and Tremblay (2012) noted that in the Tiebout–Musgrave–Oates tradition, expenditure allocation is based on the principle that national governments should be responsible for national public goods, and revenue allocation is based on the benefit principle. The idea is that redistribution should be the responsibility of the federal government and that state governments can best provide a mix of state public goods that matches the preferences of their residents. When public services are targeted to specific groups that need them, state and local policymakers can better identify those needs and design state programs accordingly. This idea stems from Oates' (1972) decentralization or subsidiarity theorem, which states: "For a public good whose consumption is defined on a geographic subset of the total population, and for which the cost of each level of public good provision is the same for the central government and the local unit concerned, local units will always be more successful (or at least as successful) in providing a pareto-efficient level of public goods in their area than the central government would be in providing a uniform level of public goods in the area of all local units." This arises simply from the fact that efficiency is increased by decision-making that is supported by better information resulting from the fact that decisions about local public goods and services require local knowledge. On the other hand, the focus on local needs makes the delivery of public services inherently complex and must be balanced against the economies of scale of centralized service delivery. The beauty of an optimal decentralization system would therefore lie in optimizing the information requirements with the economies of scale.

As Boadway and Tremblay (2012) noted, Tiebout's model was the reference point for the classical approach to fiscal federalism, but it is far from being a convincing positive description or normative prescription for fiscal federalism at the state level. Although Tiebout's stylized worldview contained many strong assumptions and was intended to be applied to the provision of local public goods, elements of it were adopted by the early literature on fiscal federalism and had a lasting influence. Among the most important were the emphasis on household mobility, the fiscal heterogeneity of different municipalities, the positive effects of fiscal competition and the taxation of services as a consequence, and the ideal of local self-government funding.

However, looking at the contributions in the literature that deal with the role of local government and the provision of local public goods and services in the context of their impact on local economic performance and development, one may well use the prism of Tiebout's methodology. The question is which local endogenous factors can lead to local economic growth. In other words, under what circumstances will businesses vote by their fit and choose favorable destination. In this context, we avoid a discussion of the economics of agglomeration, which is a natural side effect of population density in first-tier cities. We therefore focus on the opportunities for lower-tier cities to reap the benefits of local development as a consequence of their own policies and actions. Therefore, we can exclude the following factors: strategic framework, business-friendly environment, investment in infrastructure, diversity of public goods and services, education and training, taxation and tax competition, population policies. These main factors are shown in Fig. 3.1.

The role of the strategic framework of local authorities is crucial for achieving a high level of economic performance. This leads to a clear economic vision and priorities that allow for adequate resource allocation, attach great importance to the coordination of key actors, prioritize a high level of public services and investment in public infrastructure, pay attention to the labor market and education to match labor demand and supply, promote the location as a favorable business environment, etc.

A business- friendly environment influences local economic performance through the quality of local government institutions (Stankovic and Novakovic 2015) and business promotion and support for social entrepreneurship (Takano et al. 2021),

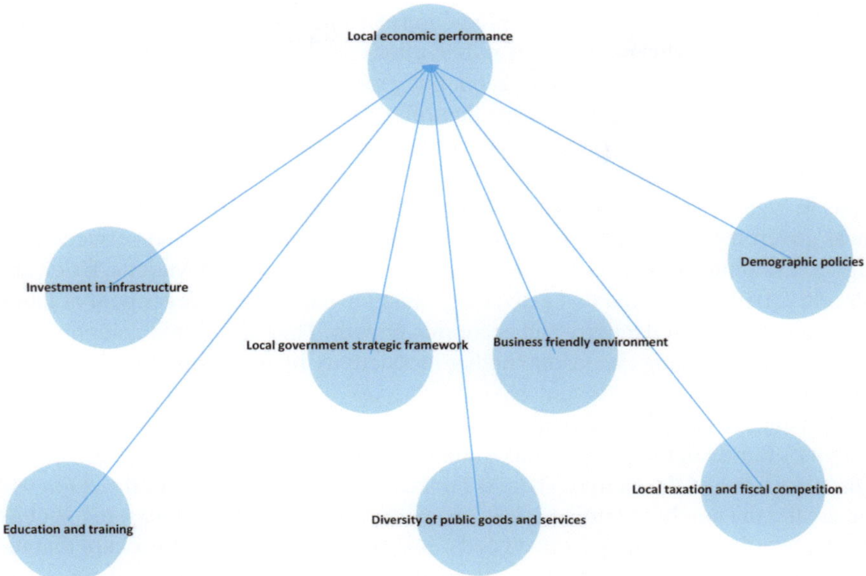

Fig. 3.1 Main factors enhancing the local economic performance. *Source* Authors

Randelovic et al. (2020). One of the most important prerequisites for a business-friendly environment is low taxes, but local governments usually have only a modest potential to benefit from tax competition between jurisdictions. On the contrary, excessive fiscal autonomy at the local level can lead to a "race to the bottom" and result in poor municipal services due to declining tax revenues. It is therefore a question of improving administrative support, removing barriers to businesses, providing full transparency and information on local potential, offering high- quality and flexible infrastructure services, etc.

Investment in infrastructure is a prerequisite for attracting businesses and increasing local economic output through the "crowd-in" effect of private investment. Collier and Venables (2016) noted that the value of infrastructure goes far beyond "user benefits," as infrastructure creates an economic environment that allows cities to exploit their potential through specialization and agglomeration effects. Similarly, Li (2017) noted that recent experience with infrastructure investment in the People's Republic of China (PRC) suggests an intertwined relationship between investment, urbanization and economic growth. One mechanism is that urbanization creates demand for infrastructure investment, which then drives economic growth through various channels, including lowering transaction costs and increasing productivity. Another mechanism highlighted in this paper is that infrastructure investment can promote urbanization by facilitating economic agglomeration toward core cities. This agglomeration process also increases productivity in the economy.

Empirical evidence supports the prediction of Tiebout's theorem that local governments that offer a diverse range of public goods and services can attract a broader range of residents, which promotes economic growth. For example, Fisher and Peters (1998) found that local governments that offer a diverse range of public goods and services have higher economic growth. They found that local governments that offer a diverse range of public goods and services have a more diverse population, which leads to higher economic growth. Similarly, Glaeser and Gottlieb (2008) found that local governments that provide a diverse range of public goods and services have higher economic growth because they generate new ideas and innovations. Hsieh and Moretti (2019) also found that a diverse population can create a larger pool of labor, leading to higher productivity and economic growth. They also found that providing education and training can lead to higher productivity by improving the skills of the labor force.

Dowding et al. (1994) argued that an impressive body of evidence suggests that Tiebout's family of models contains a number of important truths about urban policy. These are some of the claims: There is mixed but marginal support for the claim that the greater the number of jurisdictions, the greater the satisfaction with some, though not all, locally provided collective goods. It is very difficult to test the claim that the greater the number of jurisdictions, the greater the competition between them. There is virtually no significant evidence to support this assumption. There is evidence that the greater the number of jurisdictions, the lower the costs. However, this may be due to fewer opportunities for budget increases or fewer opportunities for local governments to provide certain types of public goods or redistribution remaining. The larger the number of jurisdictions, the more homogeneous they are. That this affects the efficient allocation of resources is a reasonable assumption, even if there is no solid evidence yet. An extensive empirical literature on Tiebout is warranted, and the model teaches much about the political economy of cities. For example, it specifies the microeconomic underpinnings of social and political divisions in cities, provides a political economy explanation for fluctuations in real estate values, and provides data for policy analysts and institution building.

According to Wildasin (1997), in a decentralized system, the local population can choose the combination of public goods and services and the level of local taxes to provide the desired level of public goods and services to meet their preferences. Glaeser and Gottlieb (2008) argued that local governments that offer a diverse range of public goods and services have higher economic growth thanks to the creation of new ideas and innovations. Agglomeration economies and human capital spillover effects suggest that policies of a national government aimed at strengthening the economies of specific places or regions could improve welfare. However, the mere existence of agglomeration externalities says nothing about which places should be subsidized. Without a better understanding of the nonlinearities of these externalities, it is just as likely that a government's spatial planning policies will reduce welfare as increase it. Transportation costs have historically done much to create or destroy certain places, but current transportation spending subsidizes low-income, low-density places where the effects of agglomeration are likely to be weakest.

In terms of the impact of education and training on local economic performance, there are a number of channels, such as increasing local absorptive capacity for higher value-added industries and increasing overall levels of competitiveness and innovation. Thus, the overall wage level increases when the local region becomes more productive. Interestingly, Chauvin (2018) argued that while education can lead to higher local productivity, the potential benefits to the local economy could be dampened if the educated workers migrate in search of better opportunities or if shifts in the supply of skills exceed demand growth. His research shows that education led to higher wages—mainly by allowing workers to migrate to more productive locations. For regions, however, the education program worsened wages and other labor market outcomes, but not employment, suggesting that the increased supply of educated workers outpaced demand growth. This implies that education programs must be closely aligned with current labor market needs and run in parallel with other business attraction activities. Based on a sample of 31 provinces in China between 1996 and 2010, Lv et al. (2015) found empirical evidence that the average year of schooling has a more positive impact on economic performance than capital investment and labor force. As he finds significant regional differences in the level of public education, he proposes a more supportive education policy for less developed regions, which will ultimately reduce the regional development gap.

Issues of taxation and tax competition are the subject of an extensive literature. Boadway and Bruce (1984a, b) looked at the design of a corporate tax system that is neutral with or without inflation. Business tax can relate to corporate income tax or to the taxation of the income of unincorporated businesses, e.g., within the income tax system. Neutral taxation need not depend on the information difficulties of the imputed income method or the cash flow difficulties of the current depreciation method. Mieszkowski and Zodrow (1989) examined two major issues in the literature on the Tiebout model and local taxes. Consumers of public services may be willing to pay for public goods, and at the individual community level, changes in tax rates and public services are reflected in changes in land prices and housing or rental prices. Hulten and Schwab (1991) argued that if one is to have a comprehensive income tax, and if Tiebout's model is a reasonable approximation to the structure of local government, then the implicit income from local public capital must be included in the tax base. In a pure Tieboutian world, there are no significant differences between public capital (e.g., roads, parks, schools) and private capital (e.g., residential property). Therefore, the imputed income from the street in front of the family home must be included in the comprehensive income tax base, just as the implicit income from the home itself must be included. The historical studies of Wildasin (1991) emphasize that property taxation at the local level can lead to inefficiently low amounts of local public spending. This paper shows that inefficiency can be associated with fiscal externalities: When a country increases its taxes, this leads to capital flowing to other countries, which increases their tax revenues. Inefficiency can be corrected by a subsidy that internalizes the externalities. The most important empirical parameters that determine the extent of the external effects have been determined. Illustrative calculations show that subsidy rates in the order of 40% may be necessary to achieve efficiency. John et al. (1995) Tiebout's model, which assumes that residential location choice can regulate the supply of local community goods, has attracted much criticism, but few empirical tests of its behavioral assumptions. This article presents the results of the UK's first micro study of the impact of local taxes and services on geographic mobility, a postal survey of household relocation decisions in four London boroughs during poll tax years. Taxes and services were found to be important factors in decision-making and confirmed the behavioral assumptions of the model. Respondents behaved rationally, according to Tiebout, as those who move to areas with low taxes and good services are more likely to cite low taxes and good services as driving factors than those who do the opposite. However, the policy implications still depend on political orientation. Ahmad and Brosio (2006) argued that citizens and consumers relocate to those areas that provide them with the quantity and quality of services they need at the lowest cost, especially in terms of taxation. In turn, competition ensures that the provision of services meets local preferences and minimizes inefficiencies. Although this competitive process is referred to as "voting with your feet," there is actually no policy in this model. Once we recognize that governments are not run by benevolent politicians but by interested individuals, not only is politics explicitly included in the model, but other forms of competition emerge that do not require voter mobility. Local politicians need consensus to come to or stay in power. Citizens use this quest for consensus to force local governments (politicians) to compete with each other to stay in office (Glaeser and Gottlieb 2008).

In terms of population policy, in addition to providing employment opportunities and high wages, local public services and amenities that offer a high quality of life, housing issues are becoming increasingly important. Hsieh and Moretti (2019) concluded that if there are significant differences in labor productivity between cities, output can in principle be increased by increasing employment in high-productivity cities at the expense of low-productivity cities. In spatial equilibrium, aggregate output and welfare growth are not just the sum of local shocks in each city. If workers can move between cities, a local productivity shock in one city affects wages and employment not only in that city but also in other cities. When a city with affordable housing experiences an increase in productivity, local employment increases and workers in other cities benefit from the redistribution of jobs. In contrast, if the housing supply in the city is restrictive, the redistribution of jobs is limited and the increase in productivity in the city is canceled out by higher housing costs. They conclude that local land use regulations that restrict housing supply in dynamic labor markets have significant externalities for the rest of the country. Current homeowners in highly productive cities have a private incentive to limit housing supply. By doing so, these voters are de facto limiting the number of American workers who have access to America's most productive city. In general equilibrium, this reduces the income and wealth of all American workers.

The Tiebout Theorem has important implications for local governments seeking to increase economic performance. However, we must keep in mind that Tiebout's model is, of course, not a perfect description of the real world. People are not perfectly mobile, there are not enough communities where every family can find a community with the services that suit them, etc. Moreover, contrary to the model's expectations, there are large income disparities in many communities and thus probably different desirable levels of public services (Rosen 1999). However, the contribution of the model is not negligible. Tiebout showed that the quantity and quality of the supply of public goods and services are also important in the choice of where to live (Goetz 1999). The model also shows the advantages of a decentralized supply of public goods compared to a centralized supply (Table 3.1).

3.3 Local Economic Performance and Demographic Factors

The relationship between economic performance and demographic factors and dynamics has long been the subject of debate in the field of economics. This relationship has been studied extensively by economists, demographers and sociologists, and the results have been mixed. Some studies have found a positive correlation between economic performance and demographic factors such as population growth, while others have found no significant relationship. This paper will explore the interpretation of the relationship between economic performance and demographic factors and dynamics and the various factors that may influence this relationship. The literature identifies several important channels through which demographic factors influence local economic performance: population growth, population density, age structure, education levels, income inequality, and migration levels (Fig. 3.2).

The relationship between population growth and economic performance is complex, with different studies coming to different conclusions. According to a study by Mankiw et al. (1992), population growth has a positive effect on economic performance as it increases the number of workers and the market for goods and services. A study by Bloom and Freeman (1986) also came to the conclusion that population growth has a positive impact on economic growth, as it leads to an increase in investment and technological innovation. However, other studies have found no significant correlation between population growth and economic performance. For example, a study by Lucas (1988) found that population growth has no effect on economic growth, as the increase in the working population is offset by a decrease in capital per worker. Similarly, a study by Becker and Barro (1988) found that population growth has no significant impact on economic growth, as it leads to an increase in savings but also increases the dependency ratio.

The age structure is another demographic factor that has a significant impact on economic performance. In a study by Mason (1997), it was found that the age structure of a population has a significant impact on economic growth as it affects the size of the labor force and the productivity of workers. Specifically, Mason found that a younger age structure is associated with higher economic growth, as younger workers are more productive and have a higher labor force participation rate. On the other hand, an older age structure is associated with lower economic growth, as older workers are less productive and have a lower labor force participation rate.

Table 3.1 Overview of previous research

Main conclusions	Author(s)
Introduces the model "voting by feet," where consumers (voters) move to the local community that best suits their preferences for public goods and services. A greater number of local communities means a greater difference between them, but also provides a greater opportunity for consumers to fully satisfy their preferences for public goods and services	Tiebout (1956)
He was concerned with the distribution of responsibilities for public finances. He developed the theory of fiscal federalism, in which he argued that federal systems of government are capable of solving many of the problems faced by local governments by providing the balance and stability needed to overcome problems such as the unequal distribution of wealth and the lack of commonly available resources	Musgrave (1959)
Decentralization theorem that states: "For a public good—whose consumption is defined on a geographic subset of the total population and for which the cost of each level of public good provision is the same for the central government and the local unit in question—local units will always be more successful (or at least as successful) in providing a pareto-efficient level of public goods in their area than the central government would be in providing a uniform level of public goods in the area of all local units."	Oates (1972)
This paper provides a formal model of an economy with a local public good and an endogenous jurisdictional structure (the division of a set of agents into jurisdictions) that enables the proof of Tiebout's conjecture. The results of this paper suggest that it may be possible to develop equilibrium concepts that have properties that "approximate" the optimality of the Tiebout e-equilibrium without the need for a large number of quasi- or virtual jurisdictions	Wooders (1980)
This article deals with the design of an inflation-neutral corporate tax system. The term trade tax can refer to corporation tax or to the taxation of the income of unincorporated businesses, e.g., as part of the income tax system. Neutral taxation need not depend on the information difficulties of the imputed income method or the cash flow difficulties of the current depreciation method	Boadway and Bruce (1984a, b)
This article has explored two major themes in the literature on the Tiebout model and local taxes. Consumers of public services may be willing to pay for public goods, and at the individual community level, changes in tax rates and public services will be reflected in changes in land prices and housing or rental prices	Mieszkowski and Zodrow (1989)
If we are to have a comprehensive income tax, and if Tiebout's model is a reasonable approximation of the local government structure, then the implicit income from local public capital must be included in the tax base. In a pure Tieboutian world, there are no significant differences between public capital (e.g., roads, parks, schools) and private capital (e.g., residential property). Therefore, the imputed income from the street in front of the family home must be included in the comprehensive income tax base, as well as the implicit income from the home itself	Hulten and Schwab (1991)

(continued)

Table 3.1 (continued)

Main conclusions	Author(s)
Historical studies emphasize that taxing land ownership at the local level can lead to inefficiently low levels of local public spending. This paper shows that inefficiency can be associated with fiscal externalities: When a state increases its taxes, this causes capital to flow to other states, which increases their tax revenues. Inefficiency can be corrected by a subsidy that internalizes the externalities. The most important empirical parameters that determine the extent of the external effects have been determined. Illustrative calculations show that subsidy rates in the order of 40% may be necessary to achieve efficiency	Wildasin (1991)
With constant population growth and constant capital accumulation, countries converge at about the rate predicted by the extended Solow model. They extended the Solow model to include the accumulation of human and physical capital	Mankiw et al. (1992)
An impressive body of evidence suggests that Tiebout's family of models contains a number of important truths about urban politics. These are some of the claims: There is mixed but marginal support for the claim that the greater the number of jurisdictions, the greater the satisfaction with some, though not all, locally provided collective goods. It is very difficult to test the claim that the greater the number of jurisdictions, the greater the competition between them. There is virtually no significant evidence to support this assumption. There is evidence that the greater the number of jurisdictions, the lower the costs. However, this may be due to fewer opportunities for budget increases or fewer opportunities for local governments to provide certain types of public goods or for redistribution to remain. The larger the number of jurisdictions, the more homogeneous they are. That this affects the efficient allocation of resources is a reasonable assumption, even if there is no solid evidence yet	Dowding et al. (1994)
This article presents the results of the UK's first micro study of the impact of local taxes and services on geographic mobility, a postal survey of household relocation decisions in four London boroughs during poll tax years. Taxes and services were found to be important factors in decision-making and confirmed the behavioral assumptions of the model. Respondents behaved rationally, according to Tiebout, as those who move to areas with low taxes and good services are more likely to cite low taxes and good services as driving factors than those who do the opposite. However, the policy implications still depend on political orientation	John et al. (1995)
In a decentralized system, the local population can choose the combination of public goods and services and the level of local taxes to provide the desired level of public goods and services and meet their preferences	Wildasin (1997)
Local governments that provide a wide range of public goods and services have higher economic growth. They found that local governments that offer a diverse range of public goods and services have a more diverse population, which leads to higher economic growth	Fisher and Peters (1998)

(continued)

Table 3.1 (continued)

Main conclusions	Author(s)
The main interest in fiscal federalism is the optimal distribution of functions and instruments between the central and lower levels of government, i.e., deciding which functions and instruments should be centralized and which should be left to the decentralized levels of government	Oates (1999)
Citizens and consumers relocate to those countries that offer them the quantity and quality of services they need at the lowest cost, especially in terms of taxation. In turn, competition ensures that the range of services meets local preferences and minimizes inefficiencies. Although this competitive process is referred to as "voting with your feet," there is actually no policy in this model. Once we recognize that governments are not run by benevolent policy makers but by interested individuals, not only is politics explicitly introduced into the model, but other forms of competition emerge that do not require voter mobility. Local politicians need consensus to get or stay in power. Citizens use this quest for consensus to force local governments—i.e., politicians—to compete with each other to stay in office	Ahmad and Brosio (2006)
Lower levels of government are crucial for the provision of local public goods, whose consumption is spatially restricted to the territory of subnational units, thus achieving a more efficient allocation of resources, and whose expenditure is covered by their own revenues, making them highly dependent on transfers from the central government. Therefore, it is always more efficient for local units to provide public goods and services according to the expressed preferences of citizens than for the central government to do so in the case of consumption of local public goods whose production costs are the same at the central and local levels	Oates (2005, 2006)
Voters compare the situation in their own local unit with that of other local units and punish (reward) the actions of local politicians through the public election process	Oates (2008)
Municipalities that offer a diverse range of public goods and services have higher economic growth because they generate new ideas and innovations. Agglomeration economies and spillover effects in human capital suggest that policies of a national government aimed at strengthening the economies of specific places or regions could improve prosperity. However, the mere existence of agglomeration externalities says nothing about which places should be subsidized. Without a better understanding of the nonlinearities of these externalities, it is just as likely that a government's spatial planning policy will reduce prosperity as increase it	Glaeser and Gottlieb (2008)

(continued)

Table 3.1 (continued)

Main conclusions	Author(s)
This paper assesses the significance of Tiebout's model of fiscal federalism, emphasizing its relevance to local government but questioning its applicability at the state level. It highlights Tiebout's innovative solution to the free-rider problem in public goods by proposing a model in which consumer voters reveal their preferences for public services through their choice of municipality, thereby promoting competition between local governments. This leads to an optimal distribution of budgets and a tax system that reflects the benefits of public services. Despite its influence, particularly in promoting household mobility and tax competition, the paper notes that some of Tiebout's assumptions, such as the ideal of benefit taxation at the state level, are outdated. It also recognizes the complexity of government fiscal decisions, which involve more than just economic factors and differ significantly from Tiebout's simpler community-based perspective	Boadway and Tremblay (2012)
If labor productivity varies greatly across cities, production can in principle be increased by increasing employment in high-productivity cities at the expense of low-productivity cities. In spatial equilibrium, aggregate output and welfare growth are not just the sum of local shocks in each city. If workers can move between cities, a local productivity shock in one city affects wages and employment not only in that city but also in other cities. If a city with affordable housing experiences an increase in productivity, local employment increases and workers in other cities benefit from the redistribution of jobs. On the other hand, if the city has a restrictive housing supply, the redistribution of jobs is limited and the productivity increase in the city is canceled out by higher housing costs	Hsieh and Moretti (2019)

Source Author's systematization

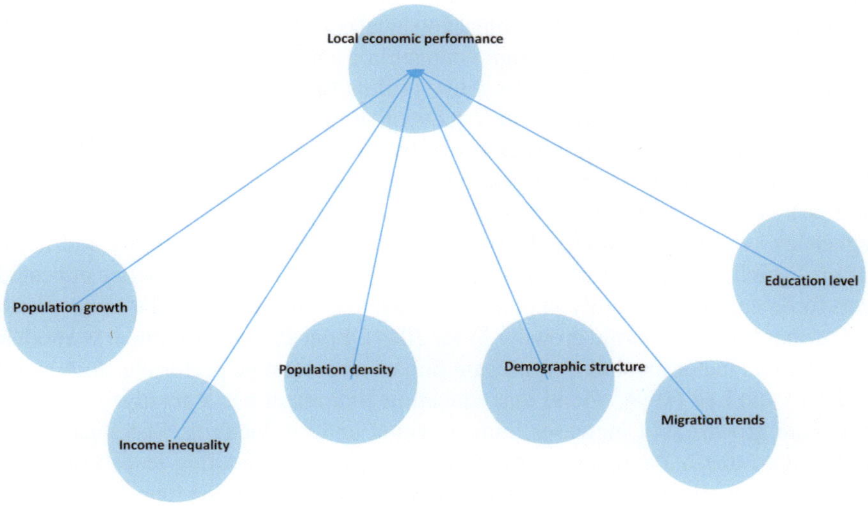

Fig. 3.2 Local economic performance and demographic factors. *Source* Authors

Chomik (2014) found that the aging of the population can have a negative impact on economic growth. The study by the Organization for Economic Cooperation and Development (OECD) found that a one percentage point increase in the proportion of the population aged 65 and over is associated with a 0.3 percentage point decline in economic growth. Aging population and local economic growth: A study by the International City/County Management Association (ICMA) found that population aging can have a negative impact on local economic growth. The study found that cities with a higher proportion of older residents tend to have lower economic growth.

Empirical evidence suggests that population aging can have a negative impact on local economic growth. A study by the International City/County Management Association (ICMA) found that cities with a higher proportion of older residents tend to have lower economic growth (ICMA 2019). However, according to a study by the Kauffman Foundation (Stangler and Litan 2014), older entrepreneurs tend to start more successful businesses than their younger counterparts. According to a study by the Milken Institute (Milken Institute 2016), the healthcare industry is likely to experience significant growth as the population ages, creating new opportunities for healthcare companies. On the other hand, the labor force participation of older workers tends to be lower than that of younger workers, which could negatively impact local economic growth if the population ages rapidly and there are not enough younger workers to replace retirees (National Bureau of Economic Research 2019). In addition, the aging population will likely increase demand for certain types of housing, such as assisted living facilities and age-restricted communities, creating new opportunities for real estate developers and builders, according to a study by Harvard College's Joint Center for Housing Studies (Joint Center for Housing Studies of Harvard College 2018). Finally, older households tend to spend less on certain types of goods and services, such as clothing and entertainment, than younger households, which could have a negative impact on local economic growth if the population ages rapidly and there are not enough younger households to replace the older ones, according to a study by the Urban Institute (Urban Institute 2017).

Several studies have shown that education is positively correlated with local economic growth. For example, a study by the National Bureau of Economic Research found that an increase in the number of college graduates in a metropolitan area is associated with higher economic growth (Moretti 2013). This is because higher levels of education are often associated with higher productivity and innovation, which can lead to higher economic output. In addition, a better educated workforce can attract highly skilled industries and businesses to the area, which can further boost economic growth (Autor 2014). Educational attainment is another demographic factor that has a significant impact on economic performance. A study by Barro and Lee (1994) found that education attainment has a positive impact on economic growth as it increases human capital and productivity. Specifically, the study found that a 1% increase in the average years of schooling leads to a 0.5% increase in economic growth. Hanushek and Woessmann (2008) found that education is positively correlated with economic growth. According to the World Bank study, an increase in the average years of schooling by one year is associated with an increase in economic growth of 0.37 percentage points.

Similarly, Ghani et al. (2014) found that higher population density is associated with higher economic growth. According to the World Bank study, a 10 percent increase in urbanization is associated with a 0.3 percentage point increase in per capita GDP growth. Population density and local economic growth: Research by the Brookings Institution suggests that population density is positively correlated with local economic growth. The study found that cities with higher population density tend to have higher economic output and productivity. The Brookings Institution found that cities with higher population density tend to have higher economic output and productivity (Berube 2016). In addition, a study by the National Bureau of Economic Research found that higher population density is associated with higher levels of innovation and entrepreneurship, which can lead to economic growth (Glaeser and Kerr 2010). However, the relationship between population density and economic growth is complex and can depend on factors such as the quality of infrastructure and the availability of skilled labor (Florida et al. 2008). In addition, high population density can lead to traffic congestion, pollution and other negative externalities that can reduce economic growth (Brueckner 2003).

In addition, the study by Hunt and Gauthier-Loiselle (2010) found that migration can have a positive effect on economic growth. A 1% increase in the proportion of foreign-born workers in a US metropolitan area is associated with a 0.14% increase in per capita income. Finally, Ostry et al. (2014) found that high levels of income inequality can have a negative impact on economic growth. An increase in the Gini coefficient (a measure of income inequality) by 10 percentage points is associated with a decline in economic growth of 0.5 percentage points. A study by the Economic Policy Institute (EPI) found that immigration can have a positive impact on local economic growth. The study found that cities with a higher proportion of immigrants tend to have higher economic output and productivity (Gould and Kierzkowski 2015). Immigrants can help fill labor shortages and bring skills and knowledge that are in demand in the local economy. In addition, immigrants are more likely to start new businesses and create jobs than natives (Kerr and Kerr 2016). However, the impact of immigration on local economic growth may depend on factors such as the skills of immigrants and local labor market conditions (Hunt and Gauthier-Loiselle 2010).

There is evidence that high levels of income inequality can have a negative impact on local economic growth. A study by the Economic Policy Institute found that cities with high levels of income inequality tend to have lower levels of economic performance and productivity (Gould and Shierholz 2011). This is because high levels of income inequality can lead to social and economic instability, which in turn discourages investment and hinders economic growth. If wealth is concentrated in a small section of the population, this can also limit the overall demand for goods and services, which can have a negative impact on local businesses (Table 3.2).

Table 3.2 Overview of previous research

Main conclusions	Author(s)
The study examines the nature and extent of the main effects of population growth on labor supply and employment in the world's developing economies (the interactions between population growth and the labor force)	Bloom and Freeman (1986)
This book shows how the interaction between technological progress and population ultimately increased the importance of education in coping with the rapidly changing technological environment, led to a significant reduction in birth rates and enabled some economies to devote greater resources to a steady increase in per capita income, paving the way for sustainable economic growth	Galor (2011)
In this paper, three models are considered and compared with the facts: a model that emphasizes the accumulation of physical capital and technological change, a model that emphasizes the accumulation of human capital through schooling, and a model that emphasizes the accumulation of specialized human capital through learning-by-doing	Lucas (1988)
This paper develops an economic analysis of the relationship between fertility rates and capital accumulation over generations. It considers the determination of fertility and capital accumulation in each generation when wage rates and interest rates are parameters for each family and for open economies	Becker and Barro (1988)
The authors conclude that the age structure of the population affects economic development more than size or growth per se and that reducing the high birth rate can create opportunities for economic growth if the right education, health and labor market policies are put in place. The report also examines specific regions of the world and how their different policy frameworks have affected the relationship between population growth and economic development	Bloom et al. (2003)
A review paper on the role of demographic change for labor force and economic growth in the APEC requires a unifying framework to organize the issues. The previous analysis has attempted to clarify the relationship between the age structure of the population, labor force participation and productivity and their combined impact on economic activity	Chomik et al. (2017)
The focus is on the challenge of encouraging people to work longer as life expectancy increases. This was the theme of a high-level political forum held recently in Brussels	OECD (2006)

(continued)

Table 3.2 (continued)

Main conclusions	Author(s)
The G20 economies have begun to address age-related challenges. The populations of the emerging G20 economies are younger than in most advanced economies, but some of them are already facing rapid demographic change at a relatively early stage of development and need to prepare for these challenges. This paper takes stock of current and projected aging societies and their impact on economic growth, productivity and inequality	Rouzet et al. (2019)
This policy snapshot outlines strategies for addressing the needs of aging urban populations and shows how internationally standardized indicators for "age-friendly cities" can be used to benchmark and predict scenarios for more informed decision-making	ICMA (2019)
This paper uses variation in the predetermined component of population aging across states to estimate the impact of aging on per capita GDP growth over the period 1980–2010. They find that every 10 percent increase in the proportion of the population over 60 reduces GDP per capita by 5.5 percent. One-third of the decline is due to slower employment growth and two-thirds to slower labor productivity growth. Their estimate implies that population aging has reduced the growth rate of GDP per capita by 0.3 percentage points per year in the period 1980–2010	Maestas et al. (2023)
Importance of a strong and robust correlation between the age structure of the workforce and the overall productivity of companies	Feyrer (2008)
This article deals with the question of how the age-related demographic change in the United States will affect entrepreneurship. There is reason to hope that, despite the problems posed by slowing population growth, entrepreneurs will find ways to capitalize on demographic trends	Stangler and Litan (2014)
Examines how the age structure of the labor force affects overall productivity and argues that the effects of aging on the dynamic channels of mobility, innovation, and entrepreneurship are more worrisome than the effects of worker age alone on firm productivity	Koettl et al. (2015)
Providing safe, affordable and accessible housing for the nation's aging population is an immediate challenge. Given the current and growing need, addressing the challenges of housing America's older adults must be a priority of the federal government	Joint Center for Housing Studies of Harvard University (2018)
It focuses on the key demographic forces that are changing the country and major metropolitan areas in the 2000s	Berube (2016)
Projection that there is a reciprocal relationship between population density and per capita income: the level of per capita income affects population growth, just as population affects per capita income via diminishing returns to labor	Olson (1996)

(continued)

Table 3.2 (continued)

Main conclusions	Author(s)
The aim is to analyze the effects of population density on development. Through established economic models such as Solow's, we know that population growth drives economic growth. The idea that higher density puts a strain on resources, makes infrastructure more vulnerable to competition, and increases the cost of providing public goods has also been discussed	Hakeem (2017)
This study provides robust evidence for these effects. They show that patent intensity, i.e., the invention rate per capita, is positively related to employment density in highly urbanized cities. A city with twice the employment density of another city has a 20 percent higher patent intensity	Carlino et al. (2007)
The rapid urbanization of the informal sector plays the strongest role, while there is some evidence of subcontracting by the formal sector and a "push" entrepreneurship. Although urbanization has continued at a rapid pace, formalization appears to have stalled. The paper also notes a modest link between this growth and increasing female labor force participation	Ghani et al. (2014)
This study shows that population density has a significant influence on economic growth. The panel regression analysis method was used to analyze the data. The analyzed data includes the population density and economic growth of selected countries in the world in the period 2011–2015	Rachmawati (2016)
This article provides an overview of empirical models of strategic interaction between governments. In order to clarify the theoretical roots of such studies, it shows how the empirical frameworks fall into two broad categories. The second part of the article addresses three econometric issues relevant to this estimation problem. A third strand of the literature deals with the analysis of strategic interactions due to "spillovers" of benefits An important strand of research in this area focuses on the choice of environmental standards by individual countries and recognizes that an interaction arises through spillover effects	Brueckner (2003)

(continued)

Table 3.2 (continued)

Main conclusions	Author(s)
This report, How Immigrants Contribute to Developing Countries' Economies, is a step forward in assessing the contribution of immigration to development and improving the design of migration and development strategies. The report empirically examines the impact of immigrants on three key components of the economy: the labor market, economic growth and public finances. It analyzes the political and historical context of immigration in each country and suggests ways to improve the contribution of immigrants in different contexts through appropriate policy measures. The report underlines the fact that the impact of immigration are not easy to identify. It depends on the context and economic conditions of the country as well as the characteristics of the immigrants. However, each country can maximize the positive impact of immigration by adopting coherent policies aimed at better managing and integrating immigrants so that they can legally invest in and contribute to the economy of the country where they work and live, while remaining secure and leading fulfilling lives	Pezzini and Tomei (2018)
Over the past two decades, the characteristics of international migration have evolved considerably and their impact has been re-examined in this paper by comparing the native-born and foreign-born in 22 OECD countries	Boubtane et al. (2014)
Immigrants in advanced economies increase production and productivity in both the short and medium terms. They find that immigrants in advanced economies increase output and productivity in both the short and medium runs. Specifically, we show that a 1 percentage point increase in the inflow of immigrants relative to total employment increases output by almost 1 percent by the fifth year	Engler et al. (2020)
They examine immigrant entrepreneurship and the survival and growth of immigrant-founded businesses compared to native businesses over time. This paper quantifies the contribution of immigrants to new business creation in a variety of areas and using several definitions. They describe differences in the types of businesses started by immigrants and their medium-term growth patterns. It also examines the relationship between these outcomes and the age of immigrants when they arrive in the United States	Kerr and Kerr (2016)

(continued)

Table 3.2 (continued)

Main conclusions	Author(s)
They measure the extent to which skilled immigrants increase innovation in the United States. The 2003 National Survey of College Graduates shows that immigrants are twice as likely to patent as natives, reflecting the fact that they disproportionately hold degrees in science and engineering. The instrument for the change in the proportion of skilled immigrants is based on the distribution of immigrants from different regions of origin across the states in 1940 and the subsequent national increase in skilled immigration from those regions	Hunt and Gauthier-Loiselle (2010)
The aim of this paper is to understand why growth outcomes vary so much. The differences in growth rates have led to dramatic differences in the average standard of living of a country's inhabitants	Barro and Lee (1994)
It has been shown that long-term data on educational attainment is collected for a large number of countries and the role of educational attainment for economic, political and social development is systematically analyzed. Moreover, educational attainment is often used as a proxy for the component of the human capital stock that is acquired in schools	Barro and Lee (1994)
This paper reviews the role of education in promoting economic well-being, focusing on the role of educational quality. It concludes that there is strong evidence that the cognitive skills of the population—rather than mere school attainment—are powerfully related to individual earnings, to the distribution of income, and to economic growth	Hanushek and Woessmann (2008)
The new economic map of America shows growing disparities, not only between people, but especially between communities. In this book, the author offers a fresh perspective on the tectonic shifts reshaping America's labor market, from globalization and income inequality to immigration and technological advancement, and how these shifts are affecting our communities	Moretti (2013)
The singular focus of public debate on the "top 1 percent" of households overlooks the component of earnings inequality that is arguably most consequential for the "other 99 percent" of citizens: the dramatic growth in the wage premium associated with higher education and cognitive ability. The central role of both the supply and demand for skills in shaping inequality, discusses why skill demands have persistently risen in industrialized countries, and considers the economic value of inequality alongside its potential social costs	Autor (2014)

(continued)

Table 3.2 (continued)

Main conclusions	Author(s)
In a relatively recent work, a new explanation for growing wage inequality has been developed, based on changes in work tasks in the economy and the resulting shifts in occupational employment	Gould and Shierholz (2011)
This paper shows how the US has suffered from rising income inequality and chronically slow growth in living standards for low- and middle-income Americans	Mishel et al. (2015)
This report shows how the US is now facing new demographic challenges	The Brookings Institution Metropolitan Policy (2010)
This paper uses a recently constructed cross-country dataset that distinguishes between market inequality and net inequality and allows us to calculate redistributive transfers for a large number of country observations. Thus, the combined direct and indirect effects of redistribution—including the growth effects of the resulting lower inequality—are on average growth-enhancing	Berg et al. (2018)
The econometric analysis, based on harmonized data from OECD countries over the last 30 years, shows that income inequality has a negative and statistically significant impact on economic growth. Most important is the gap between low-income households and the rest of the population	Cingano (2014)

Source Author's systematization

3.4 Policy Implications

The Tiebout framework provides too restrictive and outdated a framework to draw conclusions about the relationship between the provision of local goods and services and the popularity of local governments. However, theoretical and empirical advances can greatly elucidate the factors that influence the demographic and economic success of local governments. The implications of Tiebout's theorem and subsequent research underscore the importance of strategic planning, infrastructure investment, education and labor market adjustment, and a nuanced approach to fiscal decentralization in improving local economic and demographic performance. Local governments play a central role in creating an enabling environment for economic growth, innovation and community well-being through targeted policies and effective governance. There are therefore several key issues that local government should address in order to achieve positive economic outcomes:

- Establishing a strong and sound strategic framework

This framework should be based on a clear economic vision and prioritization. Local governments should develop a clear economic strategy that sets priorities, allocates resources efficiently and focuses on the sectors with the highest growth potential. This includes promoting high quality public services and investment in infrastructure, which are critical to attracting businesses and increasing economic performance. In addition, effective collaboration among key stakeholders, including businesses, educational institutions and community organizations, is critical to align economic development efforts and leverage shared strengths.

- Develop and support a business-friendly environment

Local governments should strengthen administrative support and increase transparency. Improving administrative support for businesses, reducing regulatory restrictions, and ensuring transparency around local opportunities can create a more favorable business environment and outweigh the constraints of competing solely on tax incentives. Prioritizing infrastructure development can attract businesses through the "crowd-in" effect and boost economic performance by lowering transaction costs and facilitating specialization and agglomeration effects. Promoting business and supporting social entrepreneurship can also drive local economic development while addressing social challenges.

- Aligning education and the labor market

Priority should be given to skills development and education. Investments in education and training that are aligned with labor market demand can increase local productivity and competitiveness, leading to higher wages and economic growth. However, education initiatives should be closely aligned with current labor market needs to avoid a mismatch between skills supply and demand and to ensure that the benefits of education translate into local economic gains. Attention should be focused on the labor market of the near future.

- Supporting positive demographic development

In addition to creating a dynamic labor market, there are benefits from providing a wide range of public goods and services. A wide range of public goods and services can attract a diverse population and contribute to economic growth through innovation, a larger labor pool and higher productivity. Housing affordability and land use regulations in high-productivity cities can also facilitate job redistribution, improve access to economic opportunities and increase overall economic prosperity.

Special attention must be paid to the complex relationship between demographic factors and economic performance. By understanding the different effects of population growth, age structure, education levels, population density, migration and income inequality on economic performance, policy makers can develop strategies that take advantage of these dynamics to promote sustainable economic growth. These strategies should aim to capitalize on the strengths of demographic trends while mitigating their challenges to ensure that economic development is both inclusive and robust:

- Promote labor force growth by emphasizing policies that leverage the positive impact of population growth on the labor force and market expansion. This includes supporting sectors that benefit from a larger consumer base and labor force.

- Encouraging investment in technology and innovation to capitalize on the positive impact of population growth on economic growth, as shown in studies such as Bloom and Freeman (1986).
- Implement policies that maximize the economic contribution of a younger population, such as investing in education and vocational training.
- Developing strategies to counteract the economic slowdown associated with an aging population, e.g., by incentivizing the participation of older workers and promoting sectors such as healthcare that benefit from demographic aging.
- Improve education systems to increase human capital and productivity in order to take account of the positive correlation between education attainment and economic performance, as identified by Barro and Lee (1994).
- Promote a well- educated workforce to attract high-skilled industries and capitalize on the role of education as a driver of economic growth through higher productivity and innovation.
- Supporting urban development policies that increase population density in a sustainable manner, thereby increasing economic performance and productivity.
- Promote innovation and entrepreneurship in densely populated areas, taking advantage of the positive correlation between population density and economic growth.
- Take measures to attract and integrate immigrants to harness their potential, address labor shortages, bring in in-demand skills and contribute to local economic growth through entrepreneurship.
- Tailor immigration policies to local needs and ensure that immigrants' skills match the needs of the local economy to maximize their positive impact on economic growth.
- Implement policies to reduce income inequality, recognizing its negative impact on economic stability, investment and local demand for goods and services.
- Promote inclusive economic policies that ensure that the distribution of wealth supports widespread consumption and investment, thereby supporting the local economy.

3.5 Conclusion

Attempting to implement policies that promote population and economic growth or the general well-being of local communities requires an integrated policy framework. By focusing on strategic planning, infrastructure investment, education and the nuanced management of demographic factors such as population growth, age structure and migration, local governments can promote sustainable economic growth and improve community well-being. Effective governance, stakeholder collaboration, and policies that address both economic and demographic challenges are essential to creating an enabling environment for innovation, productivity, and inclusive growth.

This literature review makes two important contributions. The first is that there is still a significant lack of comprehensive and compelling research on the non-obvious factors behind the economic and demographic outcomes of local governments. There are few both theoretical and empirical studies that address this issue. Most are single studies that examine a particular factor in a local context. Secondly, this research provides numerous prescriptions for welfare-enhancing local policies. These policy implications include both economic and demographic outcomes and their interrelationships.

There are also numerous research opportunities to consider. Suggestions can be drawn from the many policy recommendations. How can a strong and robust strategic framework for local government be designed that promotes social well-being? What are the best mechanisms to develop and support a business-friendly environment? How to design education programs that are aligned with the labor market? What are the key channels to support demographic recovery and resilience? So we can say that we still have a long way to go to close the literature gap created by Tiebout's seminal contribution.

References

Ahmad E, Brosio G (2006) Handbook of fiscal federalism. Edward Elgar Publishing Inc., USA

Autor D (2014) Skills, education, and the rise of earnings inequality among the "other 99 percent." Science 344(6186):843–851

Barro RJ, Lee JW (1994) Sources of economic growth. Carn-Roch Conf Ser Public Policy 40:1–46

Becker GS, Barro RJ (1988) A reformulation of the economic theory of fertility. Q J Econ 103(1):1–25

Berg A, Ostry JD, Tsangarides CG, Yakhshilikov Y (2018) Redistribution, inequality, and growth: new evidence. J Econ Growth 23(3):259–305

Berube A (2016) The 2016 state of metropolitan America. Brookings Institution

Bloom DE, Freeman RB (1986) The economic effects of the demographic transition. National Bureau of Economic Research

Bloom DE, Canning D, Sevilla J (2003) The demographic dividend: a new perspective on the economic consequences of population change. Rand Corporation

Boadway R, Bruce N (1984a) A general proposition on the design of a grant system. J Public Econ 24(1):75–81

Boadway R, Bruce N (1984b) A general proposition on the design of a neutral business tax. J Public Econ 24(2):231–239

Boadway R, Tremblay JF (2012) Reassessment of the Tiebout model. J Public Econ 96(11–12):1063–1078

Boubtane E, Dumont JC, Rault C (2014) Immigration and economic growth in the OECD countries, 1986–2006

Brueckner JK (2003) Strategic interaction among governments: an overview of empirical studies. Int Reg Sci Rev 26(2):175–188

Carlino GA, Chatterjee S, Hunt RM (2007) Urban density and the rate of invention. J Urban Econ 61(3):389–419

Chauvin JP (2018) Public education investment and local labor markets: evidence from a large federal program in Brazil. Discussion paper IDB-DP-606. https://publications.iadb.org/en/publications/english/viewer/Public-Education-Investment-and-Local-Labor-Markets-Evidence-from-a-Large-Federal-Program-in-Brazil.pdf
Chomik R, Piggott J, McDonald P (2017) The impact of demographic change on labour supply and economic growth: Can APEC meet the challenges ahead? https://apo.org.au/sites/default/files/resource-files/2017-02/apo-nid73910.pdf
Chomik R (2014) The impact of population ageing on economic growth: an empirical study of OECD countries.
Cingano F (2014) Trends in income inequality and its impact on economic growth. OECD social, employment and migration working papers, no. 163. OECD Publishing
Collier P, Venables AJ (2016) Urban infrastructure for development. Oxf Rev Econ Policy 32(3):391–409
Dowding K, John P, Biggs S (1994) Tiebout: a survey of the empirical literature. Urban Stud 31(4–5):767–797. https://doi.org/10.1080/00420989420080671
Engler P, MacDonald M, Piazza R, Sher G (2020) Migration to advanced economies can raise growth. international monetary fund (IMF)
Feyrer J (2008) Aggregate evidence on the link between age structure and productivity. Popul Dev Rev 34:78–99
Fisher PS, Peters AH (1998) Industrial incentives: competition among American States and Cities. WE Upjohn Institute for Employment Research, Kalamazoo, MI. https://doi.org/10.17848/9780585308401
Florida R, Mellander C, Stolarick K (2008) Inside the black box of regional development–human capital, the creative class and tolerance. J Econ Geogr 8(5):615–649
Galor O (2011) Unified growth theory. Princeton University Press
Ghani E, et al (2014) Urbanization and economic growth
Glaeser EL, Gottlieb JD (2008) The economics of place-making policies. Brookings Papers on Economic Activity, pp155–239.http://www.jstor.org/stable/27561617
Glaeser EL, Kerr WR (2010) Local industrial conditions and entrepreneurship: how much of the spatial distribution can we explain? J Econ Manag Strat 19(3):604–640
Goetz SJ (1999) Migration and local labor markets. Regional Research Institute, West Virginia University
Gould ED, Shierholz H (2011) Income inequality and the Great Recession. Economic Policy Institute
Gould ED, Kierzkowski H (2015) Immigration and economic growth in the age of the knowledge economy. Int J Bus Soc Sci 6(4):1–11
Hakeem MA (2017) Effect of population density on the level of development. Repository of Munich RePEc archive
Hanushek A, Woessmann L (2008) The role of education quality in economic growth
Hsieh CT, Moretti E (2019) Housing constraints and spatial misallocation. Am Econ J Macroecon 11(2):1–39. https://doi.org/10.1257/mac.20170388
Hulten CR, Schwab R (1991) A Haig-Simons-Tiebout comprehensive income tax. Natl Tax J 44(1). https://doi.org/10.1086/NTJ41788878
Hunt J, Gauthier-Loiselle M (2010) How much does immigration boost innovation? Am Econ J Macroecon 2(2):31–56
ICMA (2019) Aging in communities: a toolkit for local governments. International City/County Management Association
John P, Dowding K, Biggs S (1995) Residential mobility in london: a micro-level test of the behavioral assumptions of the Tiebout model. Br J Polit Sci 25(3):379–397. https://doi.org/10.1017/S0007123400007250
Joint Center for Housing Studies of Harvard University (2018) Housing America's older adults 2018. Harvard University
Kerr SP, Kerr WR (2016) Immigrant entrepreneurship. National Bureau of Economic Research

Koettl J, Sharma S, Kupets O, Mattoo A, Ozden C, Vigo JM (2015) The effects of aging on productivity: diverse, not alarming. World Bank Group
Li, Z. (2017). Infrastructure and urbanization in the People's Republic of China. ADBI working paper 632. Asian Development Bank Institute, Tokyo. https://www.adb.org/publications/infrastructure-and-urbanization-prc
Lucas RE Jr (1988) On the mechanics of economic development. J Monet Econ 22(1):3–42
Lv K et al (2015) Impacts of educational factors on economic growth in regions of China: a spatial econometric approach. Technol Econ Dev Econ 23(6):827–847
Maestas N, Mullen KJ, Powell D (2023) The effect of population aging on economic growth, the labor force, and productivity. Am Econ J Macroecon 15(2):306–332
Mankiw NG, Romer D, Weil DN (1992) A contribution to the empirics of economic growth. Q J Econ 107(2):407–437
Mason A (1997) Population, economic growth, and social welfare. World Dev 25(8):1151–1163
Mieszkowski P, Zodrow GR (1989) Taxation and the Tiebout model: the differential effects of head taxes, taxes on land rents, and property taxes. J Econ Liter 27(3):1098–1146. http://www.jstor.org/stable/2726776
Milken Institute (2016) An economic analysis of the impact of aging on the health care workforce in the United States. Milken Institute
Mishel L, Gould E, Bivens J (2015) Wage stagnation in nine charts. Economic Policy Institute
Moretti E (2013) The new geography of jobs. Houghton Mifflin Harcourt
Musgrave RA (1939) The voluntary exchange theory of public economy. Q J Econ 53(2):213–237. https://doi.org/10.2307/1882886
Musgrave RA (1959) The theory of public finance. McGraw-Hill, New York
National Bureau of Economic Research (2019) The effects of aging on the workforce and on productivity. National Bureau of Economic Research
Oates WE (1972) Fiscal federalism. Harcourt Brace Jovanovich, New York
Oates WE (1999) An essay on fiscal federalism. J Econ Liter 37(3):1120–1149
Oates WE (2005) Toward a second-generation theory of fiscal federalism. Int Tax Public Financ 12:349–373
Oates WE (2008) On the evolution of fiscal federalism: theory and institutions. Natl Tax J 61(2):313–334
Oates WE (2006) The many faces of the Tiebout model. In: Fischel W (ed.) The Tiebout model at fifty: essays in public economics in honor of wallace oates. Lincoln Institute of Land Policy, Cambidge, MA
OECD (2006) 2006 Annual report on sustainable development work in the OECD.
Olson M (1996) Distinguished lecture on economics in government: big bills left on the sidewalk: why some nations are rich, and others poor. J Econ Perspect 10(2):3–24
Ostry JD, Berg A, Tsangarides CG (2014) Redistribution, inequality, and growth. IMF staff discussion note
Pezzini M, Tomei M (2018) How immigrants contribute to developing countries' economies. OECD
Rachmawati L (2016) How can population density affect economic growth? Outlook
Ranđelović M, Nedeljković S, Jovanović M, Čabarkapa M, Stojanović V, Aleksić A, Ranđelović D (2020) Use of determination of the importance of criteria in business-friendly certification of cities as sustainable local economic development planning tool. 12:425. https://doi.org/10.3390/sym12030425
Rosen HS (1999) Public finance. Institute for Public Finance, Zagreb
Rouzet D, Sánchez AC, Renault T, Roehn O (2019) Fiscal challenges and inclusive growth in ageing societies. OECD Publishing
Samuelson PA (1954) The pure theory of public expenditure. Rev Econ Stat 36(4)
Samuelson PA (1955) Diagrammatic exposition of a theory of public expenditure. Rev Econ Stat 37(4)
Stangler D, Litan RE (2014) Age and entrepreneurship. Kauffman Foundation

Stanković J, Novaković I (2015) Impact of local self-government institutions on creating a business-friendly environment: multi-criteria analysis. In: Dolićanin Ć, Kajan E, Randjelović D, Stojanović B (eds.) Handbook of research on democratic strategies and citizen-centered E-government services. IGI Global, pp 353–373. https://doi.org/10.4018/978-1-4666-7266-6.ch018

Takano K, Ariyoshi H, Sonoda A (2021) Successful promotion of social business on the global stage: case of the City of Kitakyushu, Japan. Soc Bus 11(1–2):59–78

The Brookings Institution Metropolitan Policy (2010) State of Metropolitan America: on the front lines of demographic transformation. The Brookings Metropolitan Policy Program

Tiebout C (1956) A pure theory of local expenditures. J Polit Econ 64(5):416–424

Urban Institute (2017) The future of the US housing market: the impact of demographic trends on demand and supply. Urban Institute

Wildasin DE (1991) Interjurisdictional capital mobility: fiscal externality and a corrective subsidy. J Urban Econ 25(2):193–212

Wildasin DE (1997) Externalities and bailouts: hard and soft budget constraints in intergovernmental fiscal relations, Policy, Research working paper No. WPS 1843. World Bank, Washington DC

Wooders M (1980) The Tiebout hypothesis: near optimality in local public good economies. Econometrica 48(6):1467–1485. https://doi.org/10.2307/1912819

Marina Čolig is the Head of Finance for Zagreb County, Croatia. Her research expertise relates to fiscal equalization among local units. Since 2023, she has been the Head of the Working Group for Finance and Budget in the Croatian Union of Counties. She was a member of the Coordinating Committee for General State Adjustment during the introduction of the Euro as the official currency in the Republic of Croatia. She is also a member of the working group for risk management in the Croatian Union of Counties.

Kristina Kregar is a Teaching Assistant at the Faculty of Economics and Business, University of Rijeka, Croatia. In addition, she is a PhD researcher at Department of Economic, Business, Mathematical and Statistical Sciences, University of Trieste, Italy. Her expertise is circular economy. Kregar was a member of the organising committee of the EDT conference 2023 and 2024 and is involved in international research projects.

Saša Drezgić is a Full Professor at the Faculty of Economics, University of Rijeka, Croatia. He is an expert in public finance, fiscal policy, and investments. Prof. Drezgić has held various academic positions since 2002, including research fellowships at Cleveland State University and the London School of Economics. He served as Editor-in-Chief of Proceedings of Faculty of Economics Journal of Business and Administration. From 2013 to 2016, he directed the Center for Local Economic Development. He was also Deputy Minister of Finance of Croatia (2016-2017) and now leads the Center for Smart and Sustainable Cities (since 2018). Currently, he is the Dean of the Faculty of Economics at the University of Rijeka. Prof. Drezgić has consulted on numerous local development projects funded by international organizations and authored over 50 publications on fiscal policy and regional development.

Chapter 4
Urban Development Paths: Understanding the Intra-Regional Differentiation in the Region of Southern and Eastern Serbia

Marija Antić, Dejan Ž. Đorđević, and Danica Đurkin

Abstract At the beginning of the twenty-first century, the network of urban settlements in Serbia was characterized by uneven development. Significant differences in population concentration and degree of economic development of different categories of cities led to regional urban disparities. These disparities are probably the result of regional and local specificities, such as spatial, demographic, economic, and functional, as the region of Southern and Eastern Serbia is the most depopulated region in the country. The present work focuses on analyzing the intensity and dynamics of urban changes in the Region of Southern and Eastern Serbia and **classifying** cities according to the different urban development paths. The typological classification was performed using the K-means clustering algorithm. The cluster analysis is based on relevant demographic and socioeconomic indicators from 2002 to 2011. After the analysis, three types of cities were distinguished according to the different paths and intensity of urban development: stagnant cities with shrinkage tendencies, slight shrinkage cities, and severe shrinkage cities. The objective of the analysis is to disentangle these complex inequalities between cities and examine the impact of urban development on intra-regional differentiation. The empirical findings provide a better understanding of the heterogeneity of urban space due to the different levels of their urban development paths. In addition, the study contributes to the knowledge of intra-regional differentiation in this part of the country.

Keywords Urban development · Intra-regional differentiation · Shrinkage cities · Southern and Eastern Serbia

M. Antić (✉) · D. Đurkin
Faculty of Geography, University of Belgrade, Beograd, Serbia
e-mail: marija.antic@gef.bg.ac.rs

D. Ž. Đorđević
Faculty of Economics, University of Niš, Niš, Serbia

J. J. Stanković et al. (eds.), *Regional and Urban Challenges and Opportunities for Data-Policy Interactions*, The Urban Book Series,
https://doi.org/10.1007/978-3-032-08592-4_4

4.1 Introduction

The urban space of Serbia has always been an area of change, shaped by different cultural influences, migration flows, social regroupings, political conflicts, and economic restructuring. During the centuries, cities have experienced ups and downs, and periods of urban flourishing and prosperity have alternated with periods of stagnation and shrinkage. Thus, the specific urban identity was shaped throughout history. From the beginning of the nineteenth to the first decades of the twentieth century, the urban structure preserved the outlines of the cultural and political circles within which the cities have developed. Vojvodina cities were planned during the Austro-Hungarian period, while towns in Central Serbia represent very heterogeneous mixture of Asian, central European, and traditional Balkan attributes, which have been transforming due to spatial plans, especially since the beginning of the twentieth century. On the other side, the cities of Kosovo and Metohija and the south-western parts of Serbia, which remained part of the Turkish Empire for a long time despite planning interventions, have preserved specific characteristics from the past historical period.

Since the middle of the twentieth century, in Serbia, only 26% of the population lived in cities. Under the conditions of the new socio-political system, the process of centralized urbanization strengthened, and cities became generators of industrial development. At that time, urbanization experienced its peak, and from 1961 to 1981, the urban population grew by 1.8 million people (or 170.5%). In addition to large urban centers, significant population growth was also recorded in medium-sized and small towns. This increase in population was mainly conditioned by the relocation of the rural population and their social restructuring. The intensity of relocation to urban areas is also shown by the data on the share of immigration until 1991 when it was established that 36.5% of the urban population settled in the place of residence. Interestingly, the process of urban population growth did not lose its intensity even during the demographic transition, and its slowdown began with the "emptying" of the base of rural areas in the early 1990s (Stojanović and Vojković 2005; Đurkin et al. 2021).

Socioeconomic changes during the 1990s, marked by political and economic crises and war, changed the patterns of cities demographic development. Along with the slight increase in the total urban population from 1991 to 2011, migration flows from smaller cities to the leading regional centers (Belgrade, Novi Sad, Niš, and Kragujevac) intensified, which strengthened the polarizing effects of urbanization, spatially manifested by the deepening of the gap between different categories of cities (the capital, macro-regional centers, regional centers, sub-regional centers, and other urban settlements). This process also showed pronounced regional selectivity. Thus, significant differences in the concentration of the urban population and the degree of economic development of the cities exist (Tošić and Krunić 2005; Spasovski et al. 2012; Đurkin et al. 2021). The dominance of the Belgrade region is prominent, where 34% of the urban population lived in 16 urban settlements, according to the last census, while 29% was concentrated in the capital. On the other hand, in the Region of South and East Serbia, only 18% of the total urban population lived in 46 cities. Pronounced development disproportions are becoming more and more perceptible at the intra-regional level, and peripheral urban areas, as a rule, face demographic and economic decline, unlike those in the main development axis.

At the beginning of the twenty-first century, when more than half of the total population was concentrated in urban areas, for the first time in the last inter-census period, a drop in the urban population by 151 thousand inhabitants was recorded (3.5%). This practically means that a city the size of the macro-regional center of Niš was lost in just ten years. The data on the migration of the urban population at the national level points to a decades-long "silent" process of urban shrinkage. By 1981, only 9%, mostly small towns, showed a tendency to decrease the total population. With the slowdown in the growth of the urban population during the eighth decade of the last century, urban shrinkage became increasingly present. By 1991, a fifth of smaller urban settlements were struggling with the vortex of depopulation.

The political and economic crisis in the 1990s war years changed the direction of urban development trajectories. In the financial collapse of large production systems, which were the carriers of regional development and employment of the urban population, and the ineffective implementation of the transformation of the centralized economic structure, the former "winning cities" are becoming "losing ones." Urban growth is slowly becoming a relic of past times, and urban shrinkage, along with already deep-rooted rural shrinkage, reflects a new social reality. Only during the 1990s, more than a third of urban settlements were in the stage of urban shrinkage, and at the beginning of the twenty-first century, their share increased to 64%. According to the latest population census, almost three-quarters of cities (86%) are in the process of demographic shrinkage. The tendency of urban growth was maintained in the population, economic, and functionally most essential centers of Serbia—Belgrade and Novi Sad, in urban settlements in their immediate gravitational zone, as well as some cities in the southwestern part of Serbia with a predominantly Muslim population. The only region without a single growing town in 2022 is the Region of Southern and Eastern Serbia.

Spatial area and intensity of demographic urban shrinkage show different development dynamics. In Serbia, the most numerous (54.2%) are continuously shrinking cities, characterized by a negative average annual rate of change in the urban population (below −0.15%) during several consecutive census periods. While this group of cities is relatively evenly distributed over the entire country's territory and does not show significant inter-regional differences, their intra-regional differentiation is noticeable, and these cities are primarily located in border, peripheral, and traffic-isolated areas. Cities with episodic shrinkage are characterized by fluctuations in the total urban population development—periods of decline alternate with periods of stagnation (rarely growth) until the current phase of urban shrinkage. This group of cities (15.7%) is diverse and shows phases of two to five decades of urban shrinkage. Periods of sharp decline took place mainly during the 1990s, and after a short recovery at the beginning of the twenty-first century, they entered a shrinking phase again. According to the current demographic trends, cities recently affected by urban shrinkage (recently shrinking cities), which experienced a decline in the total population in the last inter-census period, are on the rise. It is indicative that in this group (16.3% of the total number of cities) are significant macro-regional centers, such as Niš and Kragujevac, as well as once-prosperous regional centers along development corridors (e.g. Subotica, Čačak, Kraljevo, Vranje), and these "new losers" symbolize the depth of the urban shrinkage problem at the national level. Finally, we can add a small group of cities that are in the phase of potential shrinking (potentially shrinking cities), which in the period 2011–2022 recorded mild values of the average annual rate of decline (below −0.15%), and can be considered potentially endangered, as well as the so-called "resurgent" cities, primarily located in the Belgrade region, which after one or more periods of shrinkage managed to overcome the negative phase and in the last inter-census period recorded an increase in the total population, and can carry the symbolic title of "winning cities" (Djurkin et al., 2023).

While various socio-geographical aspects of the dynamics of urban development in Serbia have been analyzed in numerous papers (e.g. Stojanović and Vojković 2005; Tošić and Krunić 2005; Grčić and Ratkaj 2006; Miletić et al. 2009; Spasovski and Šantić 2012; Göler et al. 2012; Ђурђев and Арсеновић 2015; Živanović et al. 2022), the problem of the rapidly growing phenomenon of shrinking cities, and especially the aspect that treats the regional selectivity of this process, is less present in scientific research (Antonić 2018; Živanović et al. 2021; Antonić 2022; Đurkin et al. 2021; Djurkin et al. 2023; Miljanović et al. 2023; Đurkin and Vukašinović 2023). Bearing in mind that the Region of Eastern and Southern Serbia is home to almost a third of Serbia's urban settlements and that it represents the only region in which, after several decades of urban growth, all cities have recorded a decline in the total population from 2022, this work aims to determine complex inequalities in urban development and examine the existence of intra-regional differentiation by defining cities according to the dynamics and intensity of demographic and economic development.

The research methodology is presented after the theoretical framework and analysis of the main features of urban development in the South and East Serbia Region at the beginning of the twenty-first century. The work focuses on determining different types of cities according to the dynamics and intensity of urban development and their complex analysis, as well as determining the main factors that influenced intra-regional differentiation.

4.2 Theoretical Framework

In the global context, the term development implies a change with a dual dimension—progress, which necessarily implies positive changes, and polarization, which includes spatial–temporal unevenness of development. In urban development, this term is often associated with positive change, i.e. progress (Fox and Goodfellow 2016). However, as Beauregard (1993) emphasizes, urban development is a complex process consisting of two interdependent components—urban growth and urban shrinkage, which should not be viewed dichotomously but as oppositely directed development trajectories. Bearing that in mind, urban growth is balanced by urban shrinkage in the equilibrium of urban development, as they are at opposite ends of the development "line" (Martinez-Fernandez et al. 2012).

The concept of urban growth includes the development process of cities in spatial, demographic, economic, social, and cultural terms (Fox and Goodfellow 2016). In line with this interpretation, urban practice has had a growth-oriented approach to city management for decades. On the other hand, urban shrinkage has long been neglected, even denied, due to insight through the prism of negativity (Sousa and Pinho 2015). However, this is not a new phenomenon, it is a "quiet" process that has been "in the background" of urban development for decades. Accordingly, while the issue of urban growth is deeply rooted in geographical research (e.g. Pacione 2009; Glaesr and Kahn 2004; Storper and Scott 2009; Gordon and Richardson 2011), the study of urban shrinkage has been present in the scientific literature only since the mid-1970s (van den Berg et al. 1982). It gained a broader scope by strengthening this phenomenon at the end of the twentieth century. The turning point for the global acceptance of the concept of urban shrinkage in scientific circles was the results of the project "Shrinking Cities" (Kulturstiftung des Bundes 2008).

Bearing in mind the complexity of the process itself, the components of urban development (urban growth and urban shrinkage), as well as the results of these processes (growing cities and shrinking cities), are determined and explained by observing their main aspects.

From a demographic aspect, urban growth implies increasing the city's population. In contrast, urban shrinkage refers to a city (or part of a city) experiencing a decrease in population over a short period. By the generally accepted position on the importance of the dynamics of population change, the threshold method was introduced into the literature, and increasingly applied in empirical research. The threshold method implies a pre-defined intensity of urban population change. The need for a more precise definition and "measurement" of change was realized primarily through the acceptance of the definition of shrinking cities at the international level as part of a scientific research project "Cities Regrowing Smaller" (Wiechmann and Wolf 2013). At that time, the limit value of the intensity of the total population change of -0.15% per year on average was established. With the application of a threshold value, it is possible to determine urban growth trajectory (positive average annual population change rate, higher than 0.15%), stagnation trajectory (positive average annual population change rate, lower than 0.15%), potential shrinkage trajectory(negative average annual population change rate, lower than −0.15%) and urban shrinkage trajectory (negative average annual population change rate, higher than −0,15%).

Considering the demographic aspect of urban development, the most commonly used indicator for defining and measuring the intensity of growth/shrinkage is the number of inhabitants (absolute or relative) change. Regarding the decline of the urban population in this way makes it possible to view different shrinkage continuities—from continuous and multi-year through medium and short-term shrinkage (Turok and Mykhnenko 2007; Beauregard 2009; Rink et al. 2012; Wiechmann and Wolff 2013; Martinez-Fernandez et al. 2016; Haase et al. 2016; Barreira et al. 2017; Hartt 2018; Wolff and Wiechmann 2018; Heider 2019; Barasheva et al. 2021, etc.).

The decline in the total number of inhabitants may result from negative natural growth and a lowering of the fertility rate, which occurred due to changes in the reproductive behavior of the population during rapid socioeconomic changes. The decline in fertility rates began in the mid-1960s in Western and Northern Europe, and in the 1980s, it spread to the rest of the continent. Central, Eastern, and Southeastern European countries have recorded intense fertility rate declines with socioeconomic and political changes since the 1990s (Steinführer and Haase 2007). Due to the decline in the fertility rate, the aging of population strengthens the process. On a number of examples (e.g. Italy, Germany, Poland, Japan, etc.), it has been proven that cities with a larger share of elderly are also the ones affected by the process of shrinkage (Steinführer and Haase 2007; Stryjakiewicz et al. 2012; Salone and Besana 2013; Martinez-Fernandez et al. 2016). As the reduced birth rate most often occurs due to emigration and the narrowing of the base of the young and reproductive population, the emigration of the urban population stands out as an essential component of urban shrinkage (Rieniets 2009; Turok and Mykhnenko 2007). With the intensification of city-city migrations and city-surroundings during the post-industrial phase of urban development, many cities have become emigration places of primarily young, reproductive, working, and active populations, further affecting the urban shrinkage process.

In this regard, besides the primary indicator—the decline of the total population, various indicators that explain the nature, dynamics, and intensity of the process, such as natural and migrations, are used (e.g. rate of natural increase, fertility rate, migration balance) (Stryjakiewicz et al. 2012; Hoekveld 2012; Martinez-Fernandez et al. 2016; Alves et al. 2016; Bănică et al. 2017; Wolff et al. 2017, etc.), as well as different population structures. In this context, the age structure of the population is one of the more frequently used indicators (share of different age cohorts, aging index, life expectancy, etc.) (Guimaraes et al. 2015; Martinez-Fernandez et al. 2016; Rink et al. 2012; Kabisch and Grossmann 2013; Rumpel and Slach 2014; Alves et al. 2016; Bănică et al. 2017; Chen et al. 2021).

From an economic perspective, cities characterized by urban growth are centers of increased concentration of financial activities, innovations, and drivers of social change. They can be labeled as "global winners" that have introduced many innovations in the concentration of world economic power. On the contrary, according to the opinion that urban shrinkage is a product of failed economic transformation, shrinking cities are understood as places of the so-called global "economic losers" (Hospers 2013).

Urban development took place by technological progress in the economy. The application of new technologies during the industrial phase influenced the process of urban growth to a very pronounced extent. Over time, development stagnated due to the standardization of production, which was moved to less developed countries with a cheaper production process. With deindustrialization and the sharp decline in the importance of old, "structural crisis" industrial branches (metallurgy, shipbuilding, textile, chemical industry, etc.) in the economic structure of cities, there was a dynamic urban shrinkage of the former cores of traditional industry (Friedrichs 1993).

With new services and technologies and the transition from mass production to flexible production systems, a new, post-industrial cycle of urban development begins. In the conditions of creating a networked society and restructuring global economic flows, the monetary base of cities changed. A specific group of cities grew into essential hubs in the flow of information, people and capital, the so-called "winners" of the economy at the global level, i.e. the cities characterized by growth (Fol and Cunningham-Sabot 2010; Martinez-Fernandez et al. 2012; Hospers 2013). Contrary to them, many towns whose development was based on one industry or the concentration of activities in one sector faced growing economic problems (decline in production activities, fall in gross social product, financial crisis, increase in unemployment, strengthening of emigration processes, etc.). In economic globalization, global financial flows, and internationalization of production processes, these so-called "loser" cities are becoming increasingly marginalized in attracting capital and investment. They lag behind and rapidly shrink (Martinez-Fernandez et al. 2012). For the above reasons, many authors believe that the polarization between growing and shrinking cities is a product of the post-industrial transformation of the economy and changes in the scope, structure, and sector of production (Pallagst 2009; Rieniets 2009).

Given that the transformation of the economy is considered one of the critical factors of economic growth/shrinkage in the existing literature, indicators are taken into account based on which the direction and success of the transformation of the economy and adaptation to economic changes can be seen. The most frequent indicators refer to changes in the value of the gross social product (Rink et al. 2012; Yang and Dunford 2017; Chen et al. 2021, etc.), changes in the ratio of employment and unemployment rates (Wiechmann and Pallagst 2012; Siljanoska et al. 2012, etc.). The examination of changes in the employment structure according to the dominant sectors of activity is particularly noteworthy, with a significant number of papers emphasizing the share of the working population in the production and service sectors (He et al. 2017; Wolff et al. 2017; Heider 2019; Peng et al. 2019, etc.), then changes in the education, science, and technology sector (Kabisch et al. 2012; Bănicăet al. 2017; Heider 2019), creative and knowledge-intensive business industry (Stryjakiewicz 2014), but also to the very structure of industrial production (Shan et al. 2020; Wang et al. 2020).

The extent to which the polarization between the opposite trajectories of urban development (growth and shrinkage) has deepened since the middle of the twentieth century is illustrated by the fact that between 1950 and 2000, more than 350 large European cities experienced, at least temporarily, a significant decline in the number of inhabitants (Rieniets 2009), and in the late 1990s Europe had more shrinking cities than growing ones. This group of European cities was joined by numerous post-socialist urban centers, which based their economic development on the industrial sector or even on just one industrial giant, and for this very reason, were particularly affected by the new changes and recorded a rapid urban shrinkage, both demographically, as well as economically (Fol and Cunningham-Sabot 2010). Therefore, shrinking cities represent an increasingly important phenomenon in the coming decades and represent a picture of the "demographic age of upheaval" (Rieniets 2009).

Bearing that urban development does not follow a universal pattern and that there is pronounced heterogeneity in spatial distribution, dynamics, and intensity of urban growth/shrinkage, typological classifications of cities have great practical importance in enlightening this multidimensional process. Along with the rapid strengthening of the phenomenon of shrinking cities, in the scientific literature there are more and more typologies based on the observation of the dynamics and changes in the intensity of urban shrinkage in a particular historical period (Turok and Mykhnenko 2007; Rieniets 2009; Groβmann et al. 2008; Stryjakiewicz et al. 2012; Wiechmann and Wolff 2013; Alves et al. 2016, etc.). Research that emphasizes the demographic and economic aspects of urban shrinkage and are based on the intersection of the leading indicators of these two aspects is particularly significant (Sousa 2010; Wiechmann and Pallagst 2012; Gumaraes et al. 2015; Martinez-Fernandez et al. 2016; Alves et al. 2016; Bănică et al. 2017; Wolff et al. 2017; Hartt 2018; Ribant and Chen 2019; Ma et al. 2020; Shan et al. 2020; Chen et al. 2021).

4.3 The Main Features of Urban Development in the Region of Southern and Eastern Serbia

The Eastern and Southern Serbian areas have a long and meaningful urban history. Its beginnings date back to ancient times, and after the Serbian medieval era and the period of Turkish rule, when the towns acquired recognizable outlines of traditionally Balkan attributes, the period of their flourishing began in the nineteenth century. In the new socio-political circumstances, cities along the main traffic arteries (Niš, Vranje, Leskovac, Pirot, Zaječar, Negotin, etc.) are experiencing a spatial, demographic, economic, and structural revival, thus becoming generators of urban development. Accelerated development was also encouraged in smaller cities, which are becoming increasingly essential hubs of urban life regardless of their small number and low share in the total city population. After the Second World War, a new era in the development of cities began. With urbanization, industrialization, technological innovations, and progress in traffic and communications, the urban centers of this area became the leading nuclei of general social progress. The "golden age" of city development was interrupted by the socioeconomic crisis under sanctions in the 1990s, when their economic collapse occurred, accompanied by growing economic and social upheavals. Even though their recovery was slow in the conditions of the slow post-socialist transition processes, the cities managed to preserve their demographic vitality during the following decades. However, at the beginning of the twenty-first century, demographic shrinkage gained intensity, and cities, one after another, fell into the vortex of growing depopulation. According to the 2022 population census results, all urban settlements in the Region of South and East Serbia are losing significant population shares, with four cities undergoing a metaphorical "death train" ride, losing almost a third of their population over the last three decades (Fig. 4.1).

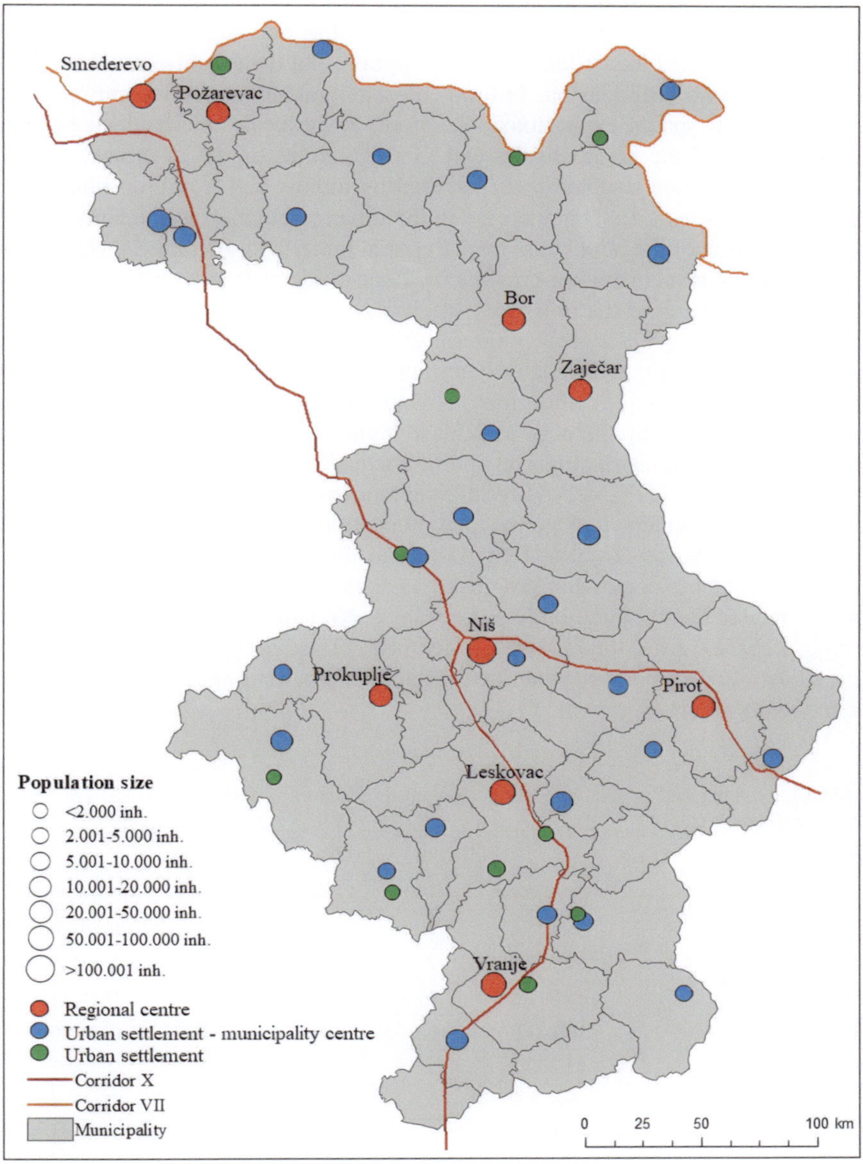

Fig. 4.1 Spatial distribution of urban settlements in the Region of Southern and Eastern Serbia—urban hierarchy and population size in 2022

Even though most of the cities in this area have a centuries-old settlement history, and some of them have always been significant urban centers in the national system of settlements (e.g. Niš, Smederevo, Pirot, Leskovac, Vranje, Zaječar), until the middle of the twentieth century only 14% of the total population of the region lived in urban areas. Their population development was only slightly accelerated until the 1960s when a more profound dynamic of urbanization began. At that time, the model of centralized industrialization was applied, and the urbanization flows took on an urban-centric dimension. In terms of demographic and economic progress, cities have vast spatial expansions. Therefore, at the peak of urbanization, from the beginning of the sixties to the end of the eighties of the twentieth century, the urban population doubled and increased by almost half a million inhabitants (Table 4.1 and Fig. 4.2). This increase in the urban population was mainly caused by the migration of the population from the rural hinterland. Attracted by the possibility of employment in the fast-growing industry, residents massively migrated from rural areas in search of the prosperity offered by cities at the time. Broader social trends, including technological progress, development of the educational system, and cultural progress, additionally contributed to the intensification of migration flows in the rural–urban relationship. As young people massively inhabited urban environments driven by the need to achieve higher aspirations, such as continuing education, professional training, and finding adequate employment, cities entered the population zenith era (Table 4.1). While the renowned centers of traditional industries and mining centers doubled the total population in just three decades (e.g. Niš, Smederevo, Leskovac, Zaječar, Bor, Aleksinac, Kostolac), it is interesting that in this period, notable population progress was experienced by smaller towns, which roughly tripled their population (e.g. Babušnica, Kuršumlija, Kladovo, Majdanpek, Svrljig, Lebane). After the relatively quick peak of socioeconomic development, a general social crisis followed when the trajectory of urban shrinkage quickly replaced the trajectory of urban growth.

Table 4.1 Number and population size of cities in the Region of Southern and Eastern Serbia, 1961, 1991, and 2022

Population size	1961		1991		2022	
	Cities	Population	Cities	Population	Cities	Population
> 100.000	–	–	1	175,649	1	178,976
50.000–100.000	1	78,712	2	177,755	3	164,306
20.000–50.000	3	85,847	6	217,894	6	183,714
10.000–20.000	6	100,613	10	148,896	7	98,356
5.000–10.000	6	47,250	11	83,412	12	86,602
2.001–5.000	21	68,667	9	31,979	9	30,974
< 2.000	9	9,957	6	6,096	8	7,024
Total	46	391,056	46	841,681	46	749,952

Source Statistical Office of the Republic of Serbia (2024)

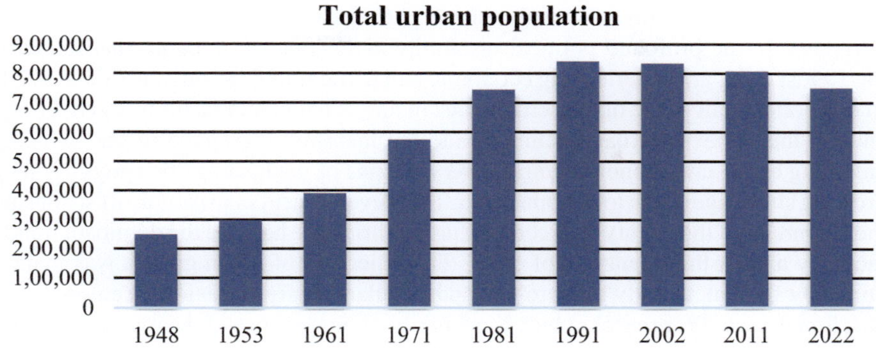

Fig. 4.2 Total urban population in the Region of Southern and Eastern Serbia, 1948–2022

Along with the disintegration of the SFR Yugoslavia, political instability, economic recession, and wars during the 1990s, slow post-socialist transition processes took place. As the structural transformation of the economic system was ineffectively implemented, the bankruptcy of the centralized economic systems that were the pillars of the region's socioeconomic development followed. The transition processes mainly affected the centers of traditional industry and mining, which were the centers of employment of the working population (Grčić and Ratkaj 2006; Miletić et al. 2009). Because the cities were not prepared for the new standards of the neoliberal market economy, in a short time, they were faced with an abundance of economic and social problems (sudden production decline, unsuccessful privatization, growth of unemployment, fictitious employment, and employment in the informal sector, strengthening of economic migrations, deepening social differences, growing urban poverty, etc.). This has had incalculable consequences for urban development, creating extreme deprivation in once-prosperous urban areas.

In the newly emerging socioeconomic circumstances, half of the cities of the South and East Serbia Region began to lose significant population share during the last decade of the twentieth century, and by 2011, the total population in 80% of urban settlements shrinkage (Table 4.1). At the same time, the mechanisms of polarization between larger and smaller cities, which until then were characterized by significantly smaller socioeconomic differences, are strengthening. The direction of migration flows is also changing, and as rural hinterlands have remained empty in previous decades, smaller cities are becoming emigration places toward larger regional centers and are experiencing more intense decline. As urban decline became the dominant development trend, in the last inter-census period the total city population documented a record decline and all cities are affected by a wave of depopulation (Table 4.1 and Fig. 4.2).

The first cities that recorded a decrease in the total population appeared from 1961 to 1991. This group of settlements was small, containing up to 13% of the total number of cities in the region. These are spa towns with a small population (e.g. Sijarinska Banja, Kuršumlijska Banja) and mining towns (e.g. Bogovina, Aleksinački

Rudnik) which, in terms of spatial and morphological aspects, had dominant rural attributes. In the profound political instability and economic collapse that marked the 1990s, cities are experiencing economic and demographic shrinkage (Table 4.2). In this period, half of the urban settlements, mainly smaller and medium-sized cities, entered the vortex of urban decline due to the inability to respond to the sudden, emerging crisis, and economic migrations were one of the leading "bad guys" of the growing shrinkage of the total population. This development matrix took off so that at the beginning of the twenty-first century, urban shrinkage became the dominant trend faced by almost three-quarters of cities. The trajectory of urban growth was maintained by the functionally most robust and population-largest regional centers (e.g. Niš, Smederevo, Požarevac), which is primarily a consequence of the immigration of the population from the surrounding, smaller urban settlements, as well as displaced persons from the area of Kosovo and Metohija. However, in the conditions of deepening regional inequalities at the national level, as well as the unsuccessful economic regeneration of once-prosperous urban centers, accompanied by the strengthening of migration flows to Belgrade and abroad, in the last inter-census period, the last bastions of urban growth in the South and East Region joined the group of shrinking cities.

We can conclude that urban shrinkage in the Region of Southern and Eastern Serbia did not have a universal character, both in terms of space and time, as well as in the manifested dynamics of the process. The temporal component of this phenomenon has been present since the 1960s, with the fact that the process of shrinkage experienced a sharp increase in the last decade of the twentieth century, since then it has been in continuous growth until 2022, when all cities enter the phase of total population decline (Table 4.2). The spatial aspect was initially limited to small towns (up to 5,000 inhabitants) that are not municipal centers, primarily located in peripheral

Table 4.2 Share of cities in the Region of South and East Serbia according to trajectories of demographic development

Trajectories of urban development	1961–1971	1971–1981	1981–1991	1991–2002	2002–2011[a]	2011–2022
Urban growth	89.1	91.3	86.9	32.6	15.5	–
Urban stagnation	–	–	–	8.7	4.4	–
Potential urban shrinkage	–	–	–	8.7	8.9	–
Urban shrinkage	10.9	8.7	13.1	50.0	71.2	100.0

Source Statistical Office of the Republic of Serbia (2024)

[a] The analysis did not include the urban settlement of Bujanovac in the periods 2002–2011 and 2011–2022. This coverage of the 2011 census was reduced due to the boycott by the majority of members of the Albanian national community.

and isolated areas. Since the 1990s, urban shrinkage has also included the category of medium-sized cities, as well as larger mining centers and industrial towns, mainly located outside the main corridor zones. Since the beginning of the twenty-first century, urban shrinkage has gradually included the leading regional centers distributed along development axes. The dynamics of this phenomenon ranged from the slow shrinkage of smaller cities until the 1990s through the pronounced intensification of this process at the end of the twentieth and the beginning of the twenty-first century, when the category of medium-sized and larger cities was particularly affected by the decline, until its escalation to all urban areas, including leading regional centers, with the fact that until 2022, individual urban settlements recorded short fluctuations in the total urban population development.

4.4 Methodology

The first methodological step includes determining the spatial and temporal framework of the research. In the region of South and East Serbia, the survey covered 46 urban settlements (27.5% of the total number of cities in Serbia), in which, according to the 2022 census, 749,952 inhabitants lived (18.2% of Serbian urban population). As it was pointed out, the classification of cities according to the trajectories of demographic development showed that with the intensification of the process of urban shrinkage, almost three-quarters of cities entered the shrinkage phase in 2002, to be joined two decades later by the remaining urban settlements of the region. For this reason, there is a need for a complex typology of urban development using the appropriate demographic and socioeconomic indicators, which would allow this multidimensional phenomenon to be seen more adequately. However, given that only demographic data from the last population census are available at the settlement level and that socioeconomic data have not yet been processed and published, a typological classification of cities according to the dynamics and intensity of urban development based on the use of demographic and socioeconomic indicators was performed for the period 2002 to 2011.

The following methodological step concerns the selection of relevant variables. Accordingly, the most commonly used indicators in the study of this phenomenon are five variables derived from the 2002 and 2011 population censuses (Statistical Office of the Republic of Serbia 2003a, 2003b, 2004, 2012, 2020, 2024), which most reliably reflect the demographic and economic component of urban development:

- the index of the total population change (2011/2002),
- the index of young population change (below 19 years) (2011/2002),
- the index of employment change (2011/2002),
- the index of employment change in secondary sector (2011/2002),
- the index of employment change in tertiary and quaternary sectors (2011/2002).

The most important methodological step refers to the typological classification of cities according to the intensity of urban development. The typological classification of cities in the Region of Southern and Eastern Serbia was performed using "K-means" cluster analysis. "K-means" is a non-hierarchical clustering method intended for grouping cities into a predetermined number of clusters. After determining the center of the cluster (centroid), the objects (cities) are sorted iteratively (by repeating the procedure), whereby they are classified based on their mutual distance from the center of the cluster. The result is mutually different clusters with relatively homogeneous intragroup characteristics (Waldrona et al. 2019). Cluster analysis represents an objective method for performing classification and has a broad scientific and practical application for determining differences in the dynamics and intensity of urban shrinkage (Sousa 2010; Hoekveld 2012; Bănică et al. 2017; Ribant and Chen 2019). By applying cluster analysis on the selected indicators, three categories of cities with different demographic and economic development dynamics were determined (stagnant cities with a shrinking tendency, slight shrinkage cities, and severe shrinkage cities).

The last methodological step was conducted to confirm the differentiation between the selected types of cities, i.e. inter-regional differences, using ANOVA analysis of variance. The analysis also made it possible to see the differences between the cities within the cluster. Implementation of "K-means" cluster and ANOVA analyses was performed using IBM SPSS Statistics 20 statistical software.

4.5 Results and Discussion

The dynamic spatial, demographic, and socioeconomic transformation of urban areas during the post-socialist transition accelerated the demographic and economic shrinkage of cities in the Region of Southern and Eastern Serbia. This is indicated by the cluster analysis results for the period from 2002 to 2011. By applying the above demographic and economic indicators, three types of cities were determined according to the dynamics and intensity of urban development—stagnant cities with a shrinking tendency, slight shrinkage cities, and severe shrinkage cities (Fig. 4.3).

Demographic specificities of selected types of cities in 2002–2011 indicate significant intra-regional disproportions. Stagnant cities with a shrinking tendency recorded a relatively stable total population development with a growing tendency to decrease the share of young people (by 9.4%). Slight shrinkage cities experienced a decrease in the total population by 7.2% and an accelerated decline in the youth contingent (17.1%). In severe shrinkage cities, a significant loss of the share of the total population (18%) was followed by the loss of almost a third of the young (Table 4.3).

The analysis of socioeconomic indicators provides a clearer insight into the intensification of the dynamics and intensity of the process of urban shrinkage. During the studied period, the category of stagnant cities with a shrinking tendency strengthened the unemployment trend. While with the collapse of the industrial giants, employment in the secondary sector fell by a third, there was a trend of increasing employment in the tertiary and quaternary sectors, with the fact that it was not the result of the post-industrial transition toward tertiary, service activities, but rather an increase in

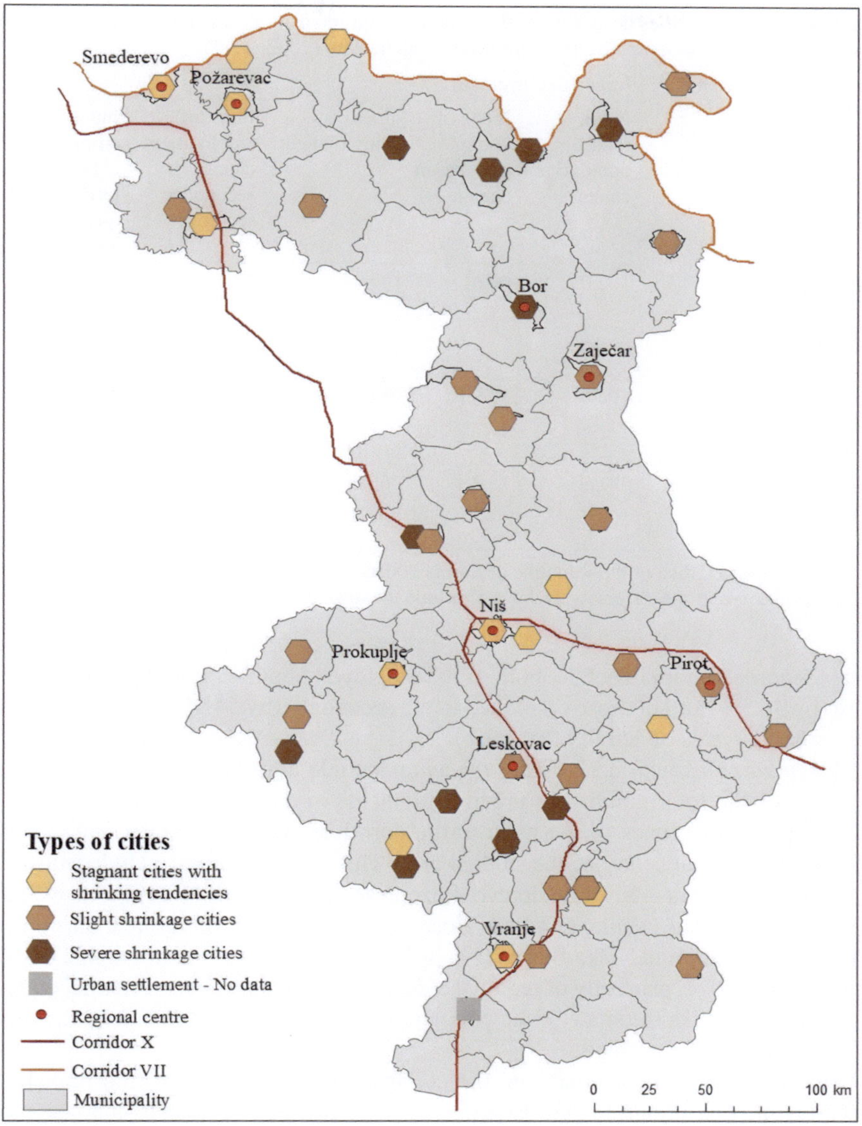

Fig. 4.3 Types of cities according to the dynamics and intensity of urban development

the number of employees in the public sector. In slight shrinkage cities, the employment index fell by 17.7% over two decades, halving the number of employees in the secondary industry, while employment in services remained almost unchanged. Severe shrinkage cities represent real "loser" cities, which demographic indicators have already indicated and socioeconomic indicators have confirmed. Namely, these

Table 4.3 Types of cities according to the dynamics and intensity of urban development—main characteristics of demographic and socioeconomic indicators (2002–2011)

Types of cities	Total number of cities[b]	Indicators (average)				
		Index of total population change	Index of young population change	Index of employment	Index of employment change in secondary sector	Index of employment change in tertiary and quaternary sectors
Stagnant cities with a shrinking tendency	13	102.0	90.6	92.1	63.7	115.7
Slight shrinkage cities	21	92.8	82.9	82.3	57.3	103.3
Severe shrinkage cities	11	81.9	71.3	63.5	45.1	95.1

Source Statistical Office of the Republic of Serbia 2003a, 2003b, 2004, 2012, 2020, 2024
[b] For the above reasons, the analysis did not include the town settlement of Bujanovac

cities recorded a rapid decline, both of the employed population in the active structure (by 36.5%) and of those employed in the secondary (by 54.9%) and tertiary and quaternary sectors of activity (by 4.9%).

Stagnant cities with a shrinking tendency include a group of 13 cities, making up 29% of the total number of urban settlements in the region. They are mainly distributed in the zones of the main traffic corridors and in the hinterland of Niš, the macro-regional center of the studied area. This group of cities includes most of the regional centers—Niš, Smederevo, Požarevac, Vranje, and Prokuplje, as well as some smaller towns (Veliko Gradište, Kostolac, Velika Plana, Babušnica, Surdulica, Svrljig, Medveđa, and Niška Banja) (Fig. 4.3). Relatively stable demographic development until 2011, primarily of regional centers (e.g. Niš, Požarevac), was accompanied by a decline in the share of the young population in the total population, which is above the cluster average.

During the economic crisis and the collapse of traditional industrial branches, once-prosperous urban centers experienced a significant decline in the share of the employed population. While the employment trend managed to maintain the macro-regional center with a relatively more diversified economy (Niš), as well as urban settlements in the mining exploitation zone (e.g. Kostolac and Požarevac), the economic monofunctional regional centers (e.g. Vranje, Prokuplje, Smederevo), as well as smaller towns whose development was closely linked to industrial giants in the nearby cities (e.g. Svrljig, Babušnica) were most affected by the recession. Although statistical data indicate a tendency for employment growth in the tertiary and quaternary sectors of activity, the cities of this group did not experience the tertiarization of the economy. Still, employment in the public sector increased sharply (the share of employees in the public sector in the total number of employees in the tertiary and quaternary sectors ranged from 37% in Kostolac, over 50% in Niš, Smederevo, and Požarevac up to 80% in Velika Plana).

The deepening of the demographic aspect of shrinkage in this category was also shown by the data of the last population census in 2022. In the previous decade, all urban settlements, including the leading regional centers, recorded a decline in the total population. The most pronounced demographic decrease took place in small towns in the southern part of the region (Surdulica, Svrljig, Niška Banja), which in the last inter-census period showed a strengthening of the population aging trend (declining the share of young people by 14%).

Slight shrinkage cities are the most numerous group of 21 urban settlements (47% of the total number of cities). They are spread over the entire region, with an increased concentration in its southern and eastern parts. Most slight shrinkage cities belong to smaller urban settlements (up to 20,000 inhabitants), except the regional centers—Leskovac, Pirot, and Zaječar (Fig. 4.3). While the aforementioned regional centers had more favorable trends in demographic development (population loss of up to 5%), the most significant decline in the total population and pronounced aging of the population were recorded by the towns with the smallest population (e.g. Boljevac, Bogovina, Petrovac na Mlavi, Vranjska Banja) and some border towns (e.g. Bosilegrad, Dimitrovrad).

Regardless of the particular more favorable demographic situation, former industrial centers (Leskovac, Zaječar, Pirot, Negotin, Kuršumlija) have a pronounced dimension of economic decline compared to the smaller towns of this group (e.g. Bogovina, Bosilegrad, Vlasotince, Kladovo). The drastic reduction in the share of the employed population (more than 20%) results from the failed post-industrial transition. In the conditions of a sharp drop in employment in secondary activities, the tertiary and quaternary sectors recorded only slight growth. At the same time, employment in the public sector is also emphasized in these cities.

This group of cities lost significant share of their population in the last intercensal period. The most important demographic shrinkage (over 15%) experienced cities with the largest population (Belo Polje, Bogovina, Vranjska Banja) and bordering, once advanced industrial centers (e.g. Zaječar, Dimitrovgrad). The accelerated strengthening of the population aging process in the mentioned depopulated cities, which lost a fifth of their young population in the period 2011–2022, is also worrying.

Severe shrinkage cities include a fifth of total region's urban settlements. This group covers peripheral, border, and monofunctional cities (industrial, mining, tourist) and small urban settlements with dominant rural attributes (Fig. 4.3). The demographic aspect of shrinkage was highly progressive, and from 2002 to 2011, in conditions of an average population decline of 18%, the contingent of the young population was reduced by almost a third. The collapse of the urban economy is accompanied by pronounced demographic shrinkage, and these so-called "losing" cities are characterized by the most intensive dynamics of urban shrinkage. The rapid decline in employment (36%) results from the halving of the number of employees in the secondary sector and the tendency to decrease employment in the tertiary and quaternary sectors (Table 4.3).

The mining centers (Bor, Majdanpek, Aleksinački Rudnik) lost a quarter of the total population in the conditions of pronounced population aging during the studied period, and the sharp drop in socioeconomic indicators is the result of the decline in the importance of this industrial branch and the lack of adequate economic regeneration of urban areas. Small spa and tourist towns with declining dominant functions suffered the most significant population losses (23% in Donji Milanovac to 34% in Sijarinska Banja). They recorded incredible values of population aging (e.g. Kuršumlijska Banja lost 79% of its young in two decades). A similar fate was experienced by small towns with predominant rural attributes (Brza Palanka, Grdelica, Vučje, Kučevo, Lebane). Bearing in mind that the results of the last census show that severe shrinkage cities in the previous decade additionally lost 16% of the total population and 23% of the contingent of the young population so that the possibility of getting out of the spiral of decline is questioned for some of them (e.g. Kuršumlijska Banja in 2022 had 57 inhabitants, and Sijarinska Banja had 327 inhabitants).

The conducted ANOVA analysis confirmed the existence of inter-regional differences between distinct clusters, which are statistically significant, as indicated by the Bonferroni test (Sig. 0.000). A statistically substantial difference also exists within the cluster, where the intra-regional differences between stagnant cities with a shrinking tendency and slight shrinkage cities are highlighted, according to the indicator of the actively employed population (Sig. 0.382), as well as according to the indicator of the employed population in the secondary sector of activity (Sig. 0.899).

4.6 Conclusion

The area of Eastern and Southern Serbia has a long and rich urban history, characterized by a wide range of development trajectories, from the "golden age" of growth to urban shrinkage. The first is the small group of cities that recorded a decreased total population from 1961 to 1991. The profound political instability and economic collapse that marked the 1990s brought half of the urban settlements of the Southern and Eastern Serbian Regions into a shrinkage trajectory. At the beginning of the twenty-first century, urban shrinkage became a dominant, unstoppable trend, covering three-quarters of cities. The results of the 2022 population census showed that the last centuries-old bastions of urban growth have "fallen" and all significant urban centers of the studied area have entered the phase of urban shrinkage (e.g. Niš, Smederevo, Pirot, Leskovac, Vranje, Zaječar). In this way, urban growth has become a relic of past times, and urban shrinkage has become a new social reality in the Region of South and East Serbia.

Urban shrinkage in spatial and temporal coverage and manifested dynamics did not have a universal path but showed pronounced intra-regional differentiation. For this reason, there is a need for a complex typology of urban development using the appropriate demographic and socioeconomic indicators, which would allow this multidimensional phenomenon to be seen more adequately. By applying cluster analysis on selected demographic and socioeconomic indicators, three categories of cities with different dynamics and intensities of urban development were determined—stagnant cities with a shrinking tendency, slight shrinkage cities, and severe shrinkage cities.

Stagnant cities with a shrinking tendency are characterized by relatively stable demographic characteristics, with a tendency for the population to age. This group mainly includes regional centers on the main traffic corridors. In the once-prosperous urban centers, the economic aspect of shrinkage is emphasized, which is reflected in the accelerated growth of the unemployment trend. Slight shrinkage cities represent the most numerous groups located mainly in the southern and eastern parts of the region. Their demographic and economic dimension of shrinkage is reflected primarily through the intensification of depopulation and the aging process, as well as the dynamic decline of employment in the secondary sector of activity. The most intense demographic and economic downturn was recorded by severe shrinkage cities, where the multi-decade population decline was followed by a rapid aging process, accentuated by a decrease in activity and employment of the population. Predominantly peripheral, border, small population, and monofunctional cities of this group represent a symbol of the growing category of "loser cities."

It is essential to note that during the last decade, all urban settlements in the Region of South and East Serbia, including the leading regional centers, have recorded population shrinkage, which has already established inter-regional differences and deepened the process of urban shrinkage. Given the complexity of the process, the presented results should be considered as a starting point for in-depth research into the phenomenon of urban shrinkage and regional polarization of urban settlements. The socioeconomic indicators of the last census will provide a clear insight into the economic aspect of urban shrinkage for 2011–2022. In addition to looking at the current trajectories of urban development, future research will focus on more detailed examination of factors of the intensity, dynamics, and differentiation of shrinking cities in the studied area.

Acknowledgements This work was supported by the European Union's Horizon Europe research and innovation program (Grant number 101059994) and by the Ministry of Science, Technological Development and Innovations of the Republic of Serbia (Contract number 451-03-65/2024-03/200091).

References

Alves D, Barreira AP, Guimarães MH, Panagopoulos T (2016) Historical trajectories of currently shrinking Portuguese cities: a typology of urban shrinkage. Cities 52:20–29

Antonić B (2018). Housing as an activator of urban development in shrinking cities: the case of cities in Vojvodina. Phd Thesis. University of Belgrade, Faculty of Architecture, National Repository of Dissertations in Serbia

Antonić B (2022) Poglavlje 8: Urbani i prostorni aspekt depopulacije u Srbiji [Chapter 8: Urban and spatial aspects of depopulation in Serbia]. In: Vuković D (ed) Nacionalni izveštaj o ljudskom razvoju—Srbija 2022—Ljudski razvoj kao odgovor na demografske promene [National Human Development Report—Serbia 2022—Human Development in response to demographic change]. UNDP, Srbija, pp 179–204

Bănică A, Istrate M, Muntele I (2017) Challenges for the resilience capacity of Romanian shrinking cities. Sustainability 9(12):2289

Barasheva E, Leng H, Barashev A, Bukhtoyarov V (2021) Typology of urban shrinkage in Russia: trajectories of Russian cities. J Urban Plan Dev 147(4):05021035

Barreira AP, Agapito D, Panagopoulos T, Guimarães MH (2017) Exploring residential satisfaction in shrinking cities: a decision-tree approach. Urban Res Pract 10(2):156–177

Beauregard RA (1993) Representing urban decline: postwar cities as narrative objects. Urban Aff Quart 29(2):187–202

Beauregard RA (2009) Urban population loss in historical perspective: United States, 1820–2000. Environ Plan A 41(3):514–528

Chen T, Hui E, Tu YW, Lang W (2021) Growth or Shrinkage? Discovering the development pattern and planning strategies for cross-border areas in China. J Urban Plan Dev 147(4):1–12

Djurkin D, Antić M, Djordjević DŽ (2023) Demographic aspects of urban shrinkage in Serbia: trajectory, variety, and drivers of shrinking cities. Sustainability 15:15961

Đurkin D, Antić M, Budović A (2021) Demographic and economic aspects of urban shrinkage in Serbia—typology and regional differentiation. Bull Serbian Geogr Soc 101(2):43–78

Đurkin D, Vukašinović S (2023) Urban sustainability in Serbia: Sustainable for whom? In: Proceedings of the International scientific conference "smart and sustainable economy: trends and perspectives. University of Niš, Faculty of Economics, Niš, pp 27–38

Fol S, Cunningham-Sabot EC (2010) Urban decline and shrinking cities: a critical assessment of approaches to urban degrowth. Ann Géogr 674(4):359–383

Fox S, Goodfellow T (2016) Cities and development, 2nd edn. Routledge, London

Friedrichs J (1993) A theory of urban decline: economy, demography and political elites. Urban Stud 30(6):907–917

Glaeser EL, Kahn ME (2004) Sprawl and urban growth. In: Handbook of regional and urban economics, vol 4. Elsevier, pp 2481–2527

Göler D, Martinović M, Ratkaj I, Šantić D (2012) Informelles Wohnen als Routine? Multiple urbane transformationen in der Agglomeration Belgrad: Das Beispiel. MittEN der Österreichischen Geogr Ges 154:212–234

Gordon P, Richardson H (2011) Urban structure and economic growth. In: Brooks N, Donaghy K, Knaap G-J (eds) The oxford handbook of urban economics and planning. Handbooks, Oxford

Grčić M, Ratkaj I (2006) Structural changes and regional differentiation of industry in Serbia during the period of transition (1988–2005). Bull Serbian Geogr Soc 86:97–112

Gumaraes MH, Barreira AP, Panagopoulos T (2015) Shrinking Cities in Portugal– where and why. Rev Portuguesa Estudos Reg 40:23–41

Haase A, Bernt M, Großmann K, Mykhnenko V, Rink D (2016) Varieties of shrinkage in European cities. Europ Urban Reg Stud 23(1):86–102

Hartt M (2018) The diversity of North American shrinking cities. Urban Stud 55(13):2946–2959

He SY, Lee J, Zhou T, Wu D (2017) Shrinking cities and resource-based economy: the economic restructuring in China's mining cities. Cities 60:75–83

Heider B (2019) What drives urban population growth and shrinkage in postsocialist East Germany? Growth Chang 50(4):1460–1486

Hoekveld JJ (2012) Time-space relations and the differences between shrinking regions. Built Environ 38(2):179–195

Hospers G (2013) Coping with shrinkage in Europe's cities and towns. Urban Des Int 18(1):78–89

Kabisch S, Grossmann K (2013) Challenges for large housing estates in light of population decline and ageing: results of a long-term survey in East Germany. Habitat Int 39:232–239

Kabisch N, Haase D, Haase A (2012) Urban population development in Europe 1991–2008—the examples of Poland and the UK. Int J Urban Reg Res 36(6):1326–1348

Kulturstiftung des Bundes (2008) Schrumpfende Städte—Shrinking Cities 2002–2008. Kulturstiftung des Bundes, FRG, Leipzig

Ma Z, Li C, Zhang J (2020) Understanding urban shrinkage from a regional perspective: case study of Northeast China. J Urban Plan Dev 146(4):1–14

Martinez-Fernandez C, Audirac I, Fol S, Cunningham-Sabot E (2012) Shrinking cities: urban challenges of globalization. Int J Urban Reg Res 36(2):213–225

Martinez-Fernandez C, Weyman T, Fol S, Audirac I, Cunningham-Sabot E, Wiechmann T, Yahagi H (2016) Shrinking cities in Australia, Japan, Europe and the USA: from a global process to local policy responses. Prog Plan 105:1–48

Miletić R, Miljanović M, Todorović M (2009) Industrial cities in transition: problem areas. Bull Serbian Geogr Soc 89(3):191–206

Miljanović D, Vuksanović-Macura Z, Doljak D (2023) Rethinking the spatial transformation of postsocialist cities: shrinking, sprawling or densifying. Cities 140:104443

Pacione M (2009) Urban geography: a global perspective, 3rd edn. Routledge, London

Pallagst K (2009) Shrinking cities in the United States of America: Three cases, three planning stories. In: Pallagst K, Aber J, Audirac I, Cunningham-Sabot E, Fol S, Martinez-Fernandez C, Moraes S, Mulligan H, Vargas-Hernandez J, Weichmann T, Wu T, Rich J (eds.) The future of shrinking cities: problems, patterns and strategies of urban transformation in a global context. University of California, Berkeley, USA, pp 81–89

Peng W, Gao W, Yuan X, Wang R, Jiang J (2019) Spatiotemporal differences in determinants of city shrinkage based on semiparametric geographically weighted regression. Sustainability 11(24):6891

Ribant M, Chen X (2019) A typology of US shrinking cities. Prof Geogr 72(1):152–164

Rieniets T (2009) Shrinking cities: causes and effects of urban population losses in the twentieth century. Nat Cult 4(3):231–254

Rink D, Haase A, Grossmann CC, Cocks M (2012) From long term shrinkage to re-growth? The urban development trajectories of Liverpool and Leipzig. Built Environ 38(2):162–178

Rumpel P, Slach O (2014) Shrinking cities in Central Europe. In: Herrschel T, Dostál P, Raška P, Koutský J (eds) Transitions in regional science—regions in transitions: Regional research in Central Europe. Wolters Kluwer, Netherlands, pp 142–155

Salone C, Besana A (2013) Urban shrinkage. Theoretical reflections and empirical evidence from a Southern European perspective. In: Proceedings of AESOP-ACSP 5th joint congress 2013: planning for resilient cities and regions. University College Dublin, Ireland, pp 1–14

Shan J, Liu J, Kong X, Liu J (2020) Identifying city shrinkage in population and city activity in the middle reaches of the Yangtze River, China. J Urban Plan Dev 146(3):04020027

Siljanoska J, Korobar V, Stefanovska J (2012) Causes, consequences and challenges of shrinkage: the case of small cities in a transition society. Built Environ 38(2):244–258

Sousa S, Pinho P (2015) Planning for shrinkage: paradox or paradigm. Eur Plan Stud 23:12–32

Sousa S (2010) Planning for shrinking cities in Portugal. Phd Thesis. Faculty of Engineering of the University of Oporto, Oporto, Portugal

Spasovski M, Šantić D, Radovanović O (2012) Historical stages in transition of natural replacement of the Serbian population. Bull Serbian Geogr Soc 92(2):23–60

Spasovski M, Šantić D (2012) Trendovi u razmeštaju i koncentraciji stanovništva Srbije—prvorazredni demografski izazov na početku XXI veka [Trends in the distribution and concentration of the population of Serbia—a first-class demographic challenge at the beginning of the 21st century]. In: Grčić M (ed) Problemi i izazovi savremene geografske nauke i nastave [The International scientific meeting Problems and challenges of contemporary geographic science and teaching]. Univerzitet u Beogradu, Geografski fakultet, Beograd, Serbia

Statistical Office of the Republic of Serbia (2003a) Census of population, households and dwellings in the Republic of Serbia 2002, age and sex, book 2, data by settlements. SORS, Belgrade, Serbia

Statistical Office of the Republic of Serbia (2003b) Census of population, households and dwellings in the Republic of Serbia 2022, activity and sex, book 5, data by settlements. SORS, Belgrade, Serbia

Statistical Office of the Republic of Serbia (2004) Census of population, households and dwellings in the Republic of Serbia 2002, industry and sex, book 6, data by settlements. SORS, Belgrade, Serbia

Statistical Office of the Republic of Serbia (2012) Census of population, households and dwellings in the Republic of Serbia 2022, age and sex, book 2, data by settlements. SORS, Belgrade, Serbia

Statistical Office of the Republic of Serbia (2020) Special statistical processing of 2011 census data—activity and sex, industry and sex, data by settlements. SORS

Statistical Office of the Republic of Serbia (2023) Census of population, households and dwellings in the Republic of Serbia 2022, Age and sex, Book 2, data by settlements. SORS, Belgrade, Serbia

Statistical Office of the Republic of Serbia (2024) Census of population, households and dwellings in the Republic of Serbia, book 20. Comparative overview of the number of population in 1948, 1953, 1961, 1971, 1981, 1991, 2002, 2011 and 2022, data by settlements. SORS, Belgrade, Serbia

Steinführer A, Haase A (2007) Demographic change as a future challenge for cities in East Central Europe. Geogr Ann B 89(2):183–195

Stojanović B, Vojković G (2005) Urbane aglomeracije na glavnim razvojnim osovinama kao polovi demografske revitalizacije Srbije [Urban agglomerations on main development areas as poles of demographic revitalization of Serbia]. Stanovništvo 43(1–4):61–80

Storper M, Scott AJ (2009) Rethinking human capital, creativity and urban growth. J Econ Geogr 9(2):147–167

Stryjakiewicz T, Ciesiolka P, Jaroszewska E (2012) Urban shrinkage and the post-socialist transformation: the case of Poland. Built Environment 38(2):196–213

Stryjakiewicz T (2014) Urban shrinkage in East-Central Europe. Bogucki Wydawnictwo

Tošić D, Krunić N (2005) Urban agglomerations in the function of regional integration of Serbia in South-Eastern Europe. Bull Serbian Geogr Soc 85(1):137–148

Turok I, Mykhnenko V (2007) The trajectory of European cities, 1960–2005. Cities 24(3):165–182

van den Berg L, Drewett R, Klaassen L, Rossi L, Vijverberg C (1982) A study of growth and decline. Pegasus, Oxford

Waldron R, O'Donoghue-Hynes B, Redmond D (2019) Emergency homeless shelter use in the Dublin region 2012–2016: utilizing a cluster analysis of administrative data. Cities 94:143–152

Wang J, Yang Z, Qian X (2020) Driving factors of urban shrinkage—examining the role of local industrial diversity. Cities 99:102646

Wiechmann T, Wolff M (2013) Urban shrinkage in a spatial perspective–operationalization of shrinking cities in Europe 1990–2010. In: Proceedings of AESOP-ACSP 5th joint congress 2013: planning for resilient cities and regions. University College Dublin, Ireland, p 932

Wiechmann T, Pallagst K (2012) Urban shrinkage in Germany and the USA: a comparison of transformation patterns and local strategies. Int J Urban Reg Res 36(2):261–280

Wolff M, Wiechmann T (2018) Urban growth and decline: Europe's shrinking cities in a comparative perspective 1990–2010. Europ Urban Reg Stud 25(2):122–139

Wolff M, Fol S, Helene R, Cunningham-Sabot EC (2017) Is planning needed? Shrinking cities in the French urban system. Town Plan Rev 88(1):131–145

Yang Z, Dunford M (2017) City shrinkage in China: scalar processes of urban and hukou population losses. Reg Stud 52(8):1–12

Živanović Z, Tošić B, Nikolić T, Samardžić I, Dabović T, Jeftić M (2021) Urban shrinkage in Serbia: the domination of economic over environmental causes. Fresenius Environ Bull 30(12):13437–13452

Živanović Z, Tošić B, Mirić N, Vračarević N (2022) The nature of urban sprawl in Western Balkan cities. Int J Urban Sci 28(2):1–20

Ђурђев Б, Арсеновић Д (2015) Градско становништво и мрежа градова [Urban population and network of cities]. In Никитовић В (ed) Популација Србије почетком 21. века [The population of Serbia at the beginning of the 21st century]. Ребуплички завод за статистику, Београд, Serbia, pp 64–70

Marija Antić is a Full Professor in the Department of Human Geography, University of Belgrade, Serbia. Her research interests are broad, and she seeks to combine empiricism and theory in understanding the urban and rural development issues of Serbia and Southeastern Europe. She has been working on a number of national and international research projects. Also, she was a team member of MIGREC – Migration, Integration and Governance Research Centre, an EU H2020 funded project. She has published more than 50 papers in peer-reviewed journals, including one textbook, two scientific monographs, and five book chapters. She participated in many conferences in Serbia and the EU. Over the last decade, she has dedicated considerable time conducting fieldwork in Serbia and Southeastern Europe.

Dejan Ž. Đorđević is a Full Professor at the Faculty of Economics, University of Niš, Serbia. He has been working on more than ten international scientific projects. He is the author and co-author of more than 70 scientific and research articles, one textbook, and one scientific monograph. His articles are published in national and international academic journals dedicated to the fields of Economic Geography, Tourism and Geography and Demography.

Danica Djurkin is a Research Assistant at the Faculty of Geography, University of Belgrade, Serbia. She is currently a PhD researcher in Geosciences. Her academic interests are in settlement geography, urban shrinkage, and population geography for urban and rural areas. Djurkin has published 27 academic and professional articles and has worked on four national and international research projects. She was a visiting researcher at the Research Centre for Territory, Transport and Environment (CITTA) at the University of Porto in October 2022.

Chapter 5
Is Return Migration the End of the Migration Cycle? Evidence from Serbia

Milica Langović, Danica Šantić, and Dejan Ž. Đorđević

Abstract Return migration is one of the most intensive migration processes in numerous countries of origin of migrants nowadays. In the Republic of Serbia, as a traditionally country of emigration, return migration has taken place in the last several decades. The aim of this paper is to examine the factors of return migration to Serbia and to investigate whether return migration is last stage of migration cycle. The paper is based on the results of an interview conducted in 2021 with return migrants in Serbia. The results show that return migration is motivated by a number of factors among which family reasons are dominant. It is also found that secondary emigration is an option for predominantly younger ones who before return in Serbia spent abroad less than five years. Understanding the migration cycle and the factors that influence return migration and secondary emigration intentions of returnees is important for migration governance, especially for a country like Serbia that is dealing with this process intensively and intend to attract migrants from abroad to return.

Keywords Return migration · Secondary emigration · Factors · Republic of Serbia

---

M. Langović (✉)
Institute of Physics Belgrade, University of Belgrade, Belgrade, Serbia
e-mail: milica.langovic@ipb.ac.rs

D. Šantić
Faculty of Geography, University of Belgrade, Belgrade, Serbia
e-mail: danica.santic@gef.bg.ac.rs

D. Ž. Đorđević
Faculty of Economy, University of Niš, Niš, Serbia
e-mail: dejan.djordjevic@eknfak.ni.ac.rs

J. J. Stanković et al. (eds.), *Regional and Urban Challenges and Opportunities for Data-Policy Interactions*, The Urban Book Series,
https://doi.org/10.1007/978-3-032-08592-4_5

5.1 Introduction

The Republic of Serbia is a country in which various migration processes take place. As far as international migration is concerned, population from Serbia has been migrating abroad since the second half of the twentieth century. At the beginning of the twenty-first century, however, other types of migration can be observed in addition to emigration. First and foremost, when it comes to the regular migration, there are return migration flows from abroad, especially from Germany and Austria, which are the most important destination countries for migrants from Serbia (Šantić et al. 2021; Stanković 2014).

Understanding the migration cycle, and why people migrate and then return to the country of origin, as well as their secondary emigration intentions, is necessary in order to formulate useful migration policy strategies. Since migration governance is responsive to the heterogeneity and complexity of migration factors and create opportunities for mutual benefits, both country and migrants, a state needs to ensure that its policy objectives are met and that systems operate with efficiency and effectiveness (Mertin et al. 2018; Taran et al. 2017; IOM 2015).

The aim of this paper is to provide insights into the factors that encourage return migration in Serbia and to investigate potentially secondary emigration intentions of returnees. So far, drivers of return migration and secondary emigration are not analyzed in details among returnees in Serbia, which gives particular significance of this paper. All the factors of return migration are determined through the prism of the period of return migration. It makes possible analysis of the evolution of factors of return migration through time framework. The research findings are important for the development of the literature on understanding the migration cycle, in the context of factors that influence the process of return migration and secondary emigration. In addition, the results of this research may be relevant for the improvement of policies and strategic documents in relation to migration governance on the different types of migration in Serbia.

5.2 Literature Review

The migration cycle is a concept that describe migration as a cycle which begins from the moment migrants decide to migrate, to their eventual return (JMDI 2015). The decision of first migration may or may not be interlinked to return migration decision. Even though some migrants planned to stay abroad for a short period of time may end up staying forever, and those with specific intentions to stay abroad permanently may return or migrate to other country (Constant 2020).

Investigating the drivers of migration represents a crucial aspect of understanding population mobility (MMC 2022). There are numerous factors behind the decision of migration. Individuals can be *pushed* out of their home countries due to poverty, economic conditions, political situation, etc., and *pulled* into destinations that offer better living conditions. So, migration commonly takes place because of the push factor that exist in the place of origin and the pull factors that exist in more developed areas. Push factors describe the reasons why individuals might emigrate from their home country, while pull factors describe the reasons why an individual might settle in a particular country (Faridi 2018; Simpson 2017; Kyaing 2012). Although migration can be driven by different factors, it is primarily considered that growing social and economic inequalities and unfulfilled life aspirations trigger the migration flows all over the world (Czaika and Reinprecht 2022).

Khan et al. (2023) in their research on migration from Bihar India to Gulf countries showed that the push factor (lack of jobs, low wages, poor financial condition, debt on the family, social insecurity, social discrimination) plays a significant role in people migrating. Apart from the push factors, some pull factors (better employment opportunities, improved living standard, personal development, presence of relatives, attractive environment) also play a significant role in. Push factors (social, political and economic) outweigh the pull factors when it comes to the migration from Northern Triangle countries (El Salvador, Honduras, and Guatemala) to the USA (NIF 2019). The results of the research conducted among the migrants living and working in Poland and Romania showed that pull factors have a greater influence on migration as compared to the push factors (Urbański 2022). Kanayo et al. (2019) examined the push and pull factors of international migration in South Africa, using Cameroonian and Congolese migrants, and showed that the mainly factor responsible for international migration are economic factors, political factors, and the influence of migrant networks.

Return migration, as an integral part of migration cycle, takes place for a variety of reasons, which are not predominantly link to economic reasons. Willingness to return relates to the reasons for emigration in the first place and the experiences of a migrants in the country of origin (Kļave and Šūpule 2019). Many migrants return for family or other private reasons, after having completed their education or work contract or having achieved another migration goals (IOM 2024). Decision to return can be also influenced by conditions in the country of destination. In some cases, when migrants fail to achieve their migration goals, or they find it difficult to integrate, return migration occurs (Matloob 2015).

Once a migrant has decided to return, there is the possibility of migrate again in the previous country of destination or in some other country (Matloob 2015). Complexity of return migration is also seen in the cases where over years return migrants move back and forth between countries of origin, transit and destination, returning briefly for visits, or envisaging a permanent return but then leaving again (Haour-Knipe and Davies 2008).

Some studies have shown that the social connection and family play a major role in return migration (OECD 2024). Using Danish administrative data from 1981 to 2005, Nikolka (2018) shows that the presence of children in the household plays a role for the return decision of migrant couples. Having children is associated with lower return propensities. The research also determined that couples are significantly more likely to return if their child is below school age. Sander (2007) explained that having spouse and children living in the country of origin yields a significant higher return migration probability. The results of research conducted among Iranian return migrants show that family ties play a central role in return migration process (Hormozinejad 2023). The author explained that return migrants highlighted the importance of family ties in their decision-making processes of return regardless of its socio-economic conditions and that family ties engender a sense of familial responsibility.

Demetradze et al. (2018) showed that the ones who have a determined plan to return have social connections to individuals in their country of origin. Niedomysl and Amcoff (2010) determined that internal return migrants in Sweden are much more likely to return for social reasons and much less likely to move for educational reasons or employment reasons. Vanthomme and Vandenheede (2021) found that return migration among various migrant groups in Belgium was highly selective in terms of older age, lower length of stay in Belgium, not living with a partner or children, being high-educated, unemployed, and in good health. Czeranowska et al. (2023) show that return and post-return plans of Polish and Lithuanian returnees are always negotiated in the context of a variety of personal, family, and professional considerations and that having a job, children, and strong attachment to the current place of living are strongest negative predictors of the likelihood of further migration.

As returnees go through several adjustment phases after their arrival in their country of origin and some of them face different challenges, some returnees require a longer period of reintegration. This means that there is a possibility that individual returnees may not fit into the community of the country of origin, which usually leads to their secondary emigration (re-migration) (Kuschminder 2017; Cortez et al. 2016; Pirvu and Axinte 2012). Therefore, a variety of different factors can influence whether reintegration takes place, including the returnee's socio-demographic characteristics, acquired skills, family and financial situation, status in society, as well as the returnee's beliefs, attitudes, and psychological characteristics. In addition to individual factors, the sustainability of return migration is also influenced by the political and socio-economic situation in the country of origin. A safe and stable environment, which includes access to resources, an adequate level of infrastructure, access to education and healthcare systems, and an environment in which human rights and needs are respected, increases the possibility of sustainable return migration (Whyte and Hirslund 2013; Black and Gent 2004).

5.3 Methodology

The analyses of the major objectives in this paper are based on the results of interviews conducted with return migrants in Serbia. Interview as a method which has proved indispensable in migration studies (Morawska 2018; Fedyuk and Zentai 2018) enabled an adequate basis to determine factors of return migration and to unveil the secondary emigration intentions of returnees.

Return migrants, as persons with extensive migration experience, who have gone through all or almost all stages of the migration cycle, represent a suitable target group for analyzing the factors of return migration and secondary emigration intentions. In order to better understand the characteristics of the respondents' migration process from various perspectives, one of the tasks of the research was to make the sample as heterogeneous as possible in terms of the socio-demographic characteristics of the respondents and the duration of their stay abroad and different period of emigration and return.

The complexity of the research process is illustrated in the form of Fig. 5.1. The interviews with the return migrants were conducted in person and by telephone. Most of the interviews were recorded and subsequently transcribed. The analysis of the interviews began by the reading of the transcribed interviews and notes. For the purposes of this paper, particular attention was paid to the following questions:

- Why did you return to Serbia; what are the main factors behind return migration?
- Do you plan to emigrate again?
- For what reasons do you plan to migrate or stay in Serbia?

The general attitudes of the interviewees were first analyzed, and then the specific common characteristics of the interviewees were singled out. The most impressive quotes related to the factors of return migration and secondary emigration intentions are highlighted. It is important to highlight that the research was anonymous and all interviewees were participated voluntarily.

In the study 43 respondents, 23 women and 20 men, took part. In terms of age structure, the majority of the sample consisted of respondents under the age of 65 (37) compared to respondents over the age of 65 (6). Most respondents were married (18) and single (10), while a smaller number live with a partner (6), divorced (5), and widowed (4). The data on the educational structure show that people with a high

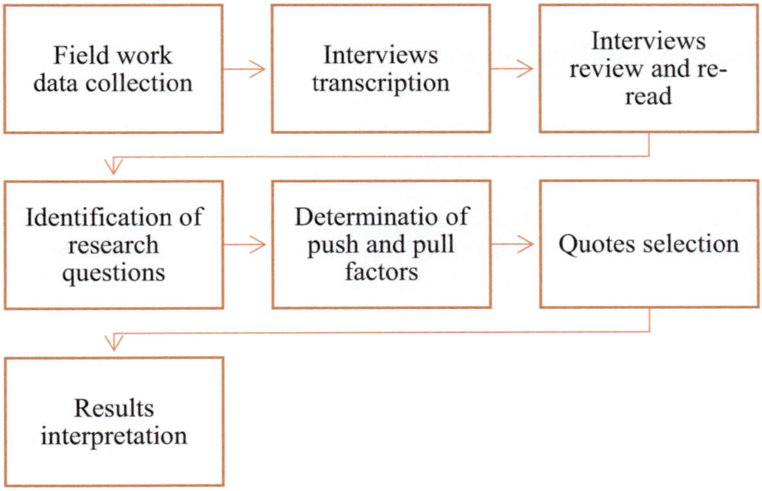

Fig. 5.1 Research process stages

Table 5.1 Socio-demographic characteristics of respondents

Characteristics		N	%
Gender	Male	20	46,5
	Female	23	53,5
Age	20–64	37	86
	65+	6	14
Marital status	Married	18	41,9
	Single	10	23,3
	Divorced	5	11,6
	Widowed	4	9,3
	Living with a partner	6	14
Education	Primary education	5	11,6
	Secondary education	21	48,8
	Tertiary education	17	39,6
Employment status	Employed	20	46,5
	Unemployed	18	41,9
	Retiree	5	11,6

Source Authors

school degree (21) and highly educated people (17) predominate, while a smaller number of respondents have a primary school education (5). Most respondents are employed (20) and unemployed (18), while the smallest number of respondents are retired (5) (Table 5.1).

5.4 Results and Discussion

Considering that return migration is direct result of emigration process, the analysis of the research provides brief insight into emigration factors in order to gain the wider picture of migration cycle characteristics.

The research results point to the numerous factors that influenced the emigration of the respondents. However, economic factors stand out as the dominant factors of emigration. It was found that respondents who emigrated in the 1970s cited financial problems, inability to find a job, and a generally poor economic situation as the main reasons for their emigration. Many of them faced poverty. The disintegration of Yugoslavia and the political situation in the country in the 1990s led to a deterioration in all areas of life and was the one of the main reasons for emigration in this period. Some interviewees explained that life in Serbia during this period was difficult for everyone, even for those who had excellent jobs. Some of the interviewees who moved out during this time (in the 1990s) were very young, in their 20s and 30s.

Push factors such as a poor economic situation and a low income and/or dissatisfaction with the job remained very important factors for emigration in the following years. In the last two decades among others, economic reasons in combination of family reasons motivate many of respondents to emigrate. Individuals who have children indicated that their decision to emigrate was largely related to the desire to provide a better life and future for their children. This is particularly related to people whose children were pupils or students at the time of emigration. Marriage abroad was also an important factor of emigration of some respondents. This was particularly emphasized by women who married abroad, mostly to people who originally came from Serbia and emigrated before them. The emigration of some interviewees was influenced by a combination of several push and pull factors. Some younger respondents who emigrated in the last decade stated that the desire to improve financial situation and desire for a new experience dominated their decision to emigrate (Table 5.2).

Return Migration. The research results show that most of the respondents have returned to Serbia in the last two decades. The duration of the interviewees' stay abroad varies. The study includes returnees who had spent one year or more than 40 years abroad before return (Table 5.2).

Talking about factors of return migration, it can be seen that in contrast to emigration, the respondents' return was primarily motivated by family reasons. It was found that family is particularly important for the return of respondents who are in their forties and fifties. Some interviewees pointed out that the main reason for their return was related to caring for elderly family members who remained in Serbia. In this context, one of the interviewees said that her mother had a serious health problem and that she had to return to take care of her. The family role in return migration process is especially important when migrants have offspring. Children have a significant influence on the process of respondents' return from abroad, mainly in the cases when they live in the country of origin. One of the interviewees explained that he returned to Serbia after divorce. He clarified that after divorce, his wife returned with the children in Serbia. He did not want to be separated from his children and decided to return shortly after them.

The education of the children is also an important factor in the return of some interviewees. In this case, the desire to educate the children in the country of origin had a dominant influence on the return of the entire family.

> I return because to the education of children. I convinced my husband that it was best for the children, so he organized our return with everything that was needed. (woman, 62)

One of the interviewees pointed out that due to the disparity between the education system in the host country and the country of origin and the general dissatisfaction with the education system in Germany, her entire family returned to Serbia.

> My biggest problem was the education of children. We returned to Serbia before the end of the school year and the child did not want to be recognized for one year of schooling in Germany, so in the new school year she had to move to a lower grade. It made me very angry. (woman, 48)

Table 5.2 Characteristic of migration cycle of respondents

N	Gender	Age	Emigration (first)	Return migration	Secondary emigration intentions
1	Female	24	2017	2019	Plan to stay
2	Female	25	2014	2021	Plan to emigrate
3	Female	25	2016	2020	Plan to emigrate
4	Female	25	2018	2019	Plan to emigrate
5	Female	28	2017	2018	Not sure
6	Female	32	2014	2016	Plan to emigrate
7	Female	37	2006	2019	Plan to stay
8	Female	38	1994	2012	Not sure
9	Female	40	2017	2019	Not sure
10	Female	42	2018	2020	Plan to emigrate
11	Female	43	1997	2015	Plane to stay
12	Female	48	2016	2018	Plan to stay
13	Female	48	2012	2020	Plan to emigrate
14	Female	48	2014	2017	Not sure
15	Female	50	2014	2019	Plan to emigrate
16	Female	50	2014	2016	Plan to stay
17	Female	56	2016	2021	Plan to emigrate
18	Female	57	2014	2020	Plan to emigrate
19	Female	62	1977	2010	Plan to emigrate
20	Female	63	1988	2014	Plan to stay
21	Female	67	1987	1991	Plan to stay
22	Female	69	1972	2018	Plan to stay
23	Female	76	1976	2013	Plan to stay
24	Male	23	2018	2019	Plan to stay
25	Male	25	2019	2019	Not sure
26	Male	29	2018	2019	Plane to emigrate
27	Male	29	1992	2006	Plan to stay
28	Male	30	2014	2018	Plan to emigrate
29	Male	31	2015	2015	Plan to stay
30	Male	31	2019	2020	Plan to emigrate
31	Male	34	2008	2012	Plan to emigrate
32	Male	37	2002	2020	Plan to stay
33	Male	39	2008	2021	Not sure
34	Male	40	1985	2009	Plan to stay
35	Male	46	2013	2018	Plan to emigrate
36	Male	48	1994	2014	Plan to stay

(continued)

Table 5.2 (continued)

N	Gender	Age	Emigration (first)	Return migration	Secondary emigration intentions
37	Male	49	1995	2012	Plan to emigrate
38	Male	50–60	2019	2020	Plan to emigrate
39	Male	52	2006	2021	Plan to stay
40	Male	57	1989	2013	Plan to stay
41	Male	70	1971	2007	Plan to stay
42	Male	71	1970	2016	Plan to stay
43	Male	73	1970	2008	Plan to stay

Source Authors

The end of the employment contract is also one of the most important factors that encourage the return of the whole family. Regardless of whether the respondents desire to stay in the country of destination or not, they were obliged in some way to return to Serbia.

> My husband had a work contract for 4 years, after the end of that period, we return to Serbia. (woman, 67)

Among the younger interviewees (younger than 30 years old), there are those who returned with the desire to finish their education in Serbia, but also because of family. One of the interviewees noted that it is much easier to study in her native language.

Longing for family in the combination with the opportunity to get a job in the country of origin were the main factors of return of one of the respondents.

> I returned to Serbia because my father found me a job in my hometown. But at the same time, I missed my father, sister, friends. I missed Serbia. You know, everything was great in Montenegro, but somehow, we felt it was time to return and we did it. (man, 34)

It was found that pensioners return primarily with the desire to spend their retirement in Serbia. In addition, some stated that they were return for *the people and the language*. Long-term stays abroad have influenced the fact that they enjoy a well-deserved pension after their arrival in Serbia.

> I wanted to return to my homeland, you got bored of that foreign country at some point, I wanted to hear our birds and our language, I got tired of German, I didn't want to die there. (man, 70)

Among the older returnees, there is also a tendency for the return to take place after one of the partners has been initiated to return. One of the respondents stated that she returned because her husband wanted so. She stated that her husband had a desire to return for a long time, but she postponed it because she wanted to stay with the children abroad. However, she decided to support her husband and return to Serbia. Both of them are retired.

Although the results of the study show that pull factors play the dominant role in the return process, it was found that in some cases the negative characteristics of the integration process in the host country can contribute to the return of migrants to Serbia. Though labor market conditions in Western European countries are much more favorable than in Serbia, the expectations of some respondents were higher.

> I returned because the salary was too low. I knew that I could earn almost the same money if I did a similar job in Serbia… I felt a little bit disappointed because I expected (and was told) one and got another. (man, 31)

Changes of living conditions and some ecological conditions in the countries of destination can also motivate migrants to the return. In this regard, some of the respondents explained that various push factors contributed their return.

> My decision to return to Serbia was very sudden. I impulsively decided to return. After so many years of living there, their system started to change, it became more stressful and harder to live there, it became very polluted and I no longer felt welcome. My standard of living started to decline and then I decided to return to Serbia. (man, 40)

> I wasn't happy, I had had enough of everything. Everything started to bother me, from the climatic conditions to the job itself. (woman, 25)

One of the reasons of return of some respondents was Covid-19 pandemic. The impact of the pandemic was felt on an international, national and even individual level, and priorities of people worldwide changed overnight. The political and economic systems of countries around the world took a hit and influenced flows of return migration (Rayan and Arcand 2023).

What After Return? Looking at the results of the survey with regard to the respondents' future migration plans shows that in the survey sample there are those who plan to stay in Serbia (20 respondents), those who plan to emigrate (17 respondents), and those who are not sure/have not yet decided (6 respondents) (Table 5.2). Therefore, it can be concluded that return migration is not exclusively last stage of the migration cycle. Certain differences were established between those who plan to move again and those who plan to stay in Serbia.

It was found that respondents who have spent several years outside the countries are being the most likely to consider secondary emigration. The results show that respondents want to move abroad primarily because of higher salaries and better incomes. Many of them have not managed to integrate into the labor market and are not satisfied with the living conditions in Serbia after their return. Considering that these are people who have already had the opportunity to live abroad, who are familiar with current events and the benefits of living in one country and the other, and who can make a decision about a second emigration based on their personal preferences.

> I plan to leave Serbia again … This time I would try to move to America… salaries are excellent in all sectors there… I would do any job in USA. Of course, I would like the most to find a job in the field of hospitality or tourism, but I am ready to do any other job for the sake of going to America. (woman, 25)

Among the respondents who have not yet decided whether they will leave or stay, there are mainly those who stated that their migration intentions depend on their partner's decision to emigrate (e.g., that they will leave if their partner gets a job abroad) or that they will leave if there is an opportunity for a better paid job abroad, or they will stay if they find an adequate job in Serbia, etc. One of the interviewees stated that she became pregnant after returning to Serbia and that she is now focusing on the pregnancy and not thinking about emigrating.

> I am not sure if I would migrate again. In fact, I don't think about it now. All I think about right now is my pregnancy and baby… I live day by day. I will not migrate at least until the baby grows up, and then we will see. (woman, 28)

Respondents who have a job and are satisfied with their salary and who have a family and especially children in Serbia usually do not plan to migrate again. Among those who plan to stay in Serbia, there are also those who are over 60 years old. Predominantly, those respondents spent abroad more than 30 years. Some of them explained that they returned to Serbia in order to spend their retirement in the country of origin. It was found that they are generally satisfied with their living conditions in Serbia and that they are financially secure because of their foreign pension. During their stay abroad, they frequently visited their country of origin and were in contact with the local population in Serbia. This suggests that they were transnationally active. After return, most of them have a family abroad and occasionally visited them, but their plans when it comes to emigration intentions are not related anymore in living abroad.

5.5 Conclusion

This article analyzes the factors of return migration and secondary emigration intentions of returnees in Serbia. In order to achieve main objectives of the research interviews were conducted with return migrants. The importance of the interview is demonstrated by the fact that it enabled a deeper understanding of the decisions and motives that led the person to first return to their country of origin and potentially to emigrate again.

International return migration has received increase attention due to the fact that many countries are facing the intensifying numbers of returnees (Ndreka 2019). For Serbia, as a country facing negative demographic characteristics and decades of population decline, the return of people from abroad can be an important factor in improving demographic conditions but also socio-economic situation, due to returnee's human, social, and financial capital. The results of the study show that family reasons, such as caring for parents and/or children in Serbia, return at the initiation of the partner/children were particularly prominent when it came to the return of the interviewees. Although family reasons dominated in most cases, it is important to point out that the role of the family in return can be combined with some other factors. It appears that entire families return from abroad because of the children and their needs. Although official data indicate that a significant proportion of returnees in Serbia are pensioners, the fact that younger people are also returning to Serbia is encouraging. In addition to family reasons, which are an important pull factor in the return process, it was found that some return due to push factors, i.e., due to certain limitations in the field of social and economic integration in the host country, which is especially the case for younger returnees. Some other externa factors such as pandemic Covid-19 influenced return of some returnees.

The process of emigration and the process of return migration are two sides of the same coin, i.e., return migration depends on the process of emigration. The emigration is a process that has a negative impact on the overall development of the country of origin. The effects of this process are particularly noticeable when the young population leaves the country. Serbia directly lost the population who emigrated, but indirectly also their children, if they left together and/or were born in another, foreign country. In addition to the consequences for the number of inhabitants, emigration also had an impact on the lack of births and the aging of the population, which are a reality in Serbia. So, in the Republic of Serbia, emigration and negative natural growth contribute to the decades-long process of population decline. In the foreseeable future, Serbia, as a country of emigration, will continue to be exposed to various undesirable effects (Arandarenko 2021; Rašević 2016). Emigration is not only a negative feature of demographic development, but also of the country's overall development. When the population leaves an area, human and social capital is lost, which is a necessary component of a country's overall development.

Therefore, the migration policy of the country of origin should primarily focus on reducing emigration. As it turns out that push factors dominate emigration, the policy should primarily aim at creating conditions that enable young people to find a job and a decent salary in the country of origin. Since the role of the family in the return process has been emphasized, it is necessary to work on ensuring the conditions for the return of migrants from abroad to Serbia to a greater extent. Since it is the young returnees who can contribute to the overall development of the country of origin, the focus should be on developing measures that can promote their return. The results have confirmed that the return of migrants to their country of origin does not mean that they will stay there for the rest of their lives and that many of the interviewees plan to move abroad again. Secondary emigration is particularly common among respondents who are not satisfied with living conditions in Serbia.

Acknowledgements The data used in this article was collected during the implementation of the research project "Trajectories of Reintegration. The Impact of displacement, Migration and Return on Social Change" (2019–2022) which has been supported by the German Federal Ministry for Economic Cooperation and Development (BMZ).

References

Arandarenko M (2021) Nacionalna studija o Srbiji: Razumevanje interakcije izmedju migracije, ljudskog kapitala I tržišta rada. European Training Foundation

Black R, Gent S (2004) Defining, measuring and influencing sustainable return: the case of the Balkans. Working paper T7. Sussex Centre for Migration Research

Constant AF (2020) Time-space dynamics of return and circular migration: theories and evidence, GLO Discussion Paper, 446, Global Labor Organization (GLO), Essen

Cortez M, Rosario C, Dino MJ (2016) Eliciting challenges on social connectedness among Filipino nurse returnees: a cross-sectional mixed-method research. Nursing Res Pract

Czaika M, Reinprecht C (2022) Migration drivers: why do people migrate?. In: Scholten P (eds) Introduction to migration studies. IMISCOE Research Series. Springer, Cham. https://doi.org/10.1007/978-3-030-92377-8_3

Czeranowska O, Parutis V, Trąbka A (2023) Between settlement, double return and re-emigration: motivations for future mobility of Polish and Lithuanian return migrants. Comp Migr Stud CMS 11(28). https://doi.org/10.1186/s40878-023-00350-3

Demetradze T, Dvarionaitė J, Ivinskytė I (2018) Factors affecting return migration. LCC International University. https://doi.org/10.13140/RG.2.2.31260.92805

Faridi R (2018) Migration theories: lee's push pull theory. https://rashidfaridi.com/2018/04/05/migration-theories-lees-push-pull-theory/

Fedyuk O, Zentai V (2018) The interview in migration studies: a step towards a dialogue and knowledge co-production? In: Zapata-Barrero R, Yalaz E (eds) Qualitative research in European migration studies. IMISCOE Research Series. https://doi.org/10.1007/978-3-319-76861-8_10

Haour-Knipe M, Davies A (2008) Return migration of nurses. International Centre on Nurse Migration

Hormozinejad S (2023) 'Why did you return?': North–South Return migration and family ties in the case of Iran. Sozialpolitik. Ch, 2, pp 2–19

IOM (2015) Migration governance framework. https://www.iom.int/sites/g/files/tmzbdl486/files/2019-01/S-16-9-Migration-Governance-Framework.pdf

IOM (2024) IOM's policy on the full spectrum of return, readmission and reintegration. https://www.iom.int/sites/g/files/tmzbdl486/files/our_work/DMM/AVRR/ioms-policy-full-spectrum-of-return-readmission-and-reintegration.pdf

JMDI (2015) Managing the link between migration and local development. Joint Migration and Development Initiative

Kanayo O, Anjofui P, Stiegler N (2019) Push and pull factors of international migration: evidence from migrants in South Africa. J Afr Union Stud (JoAUS) 8(2):219–250. https://hdl.handle.net/10520/EJC-1830c2107f

Khan MI, Alharthi M, Haque A, Illiyan A (2023) Statistical analysis of push and pull factors of migration: a case study of India. J King Saud Univ Sci 35(8). https://doi.org/10.1016/j.jksus.2023.102859

Kļave E, Šūpule I (2019) Return migration process in policy and practice. In: Kaša R, Mieriņa I (eds) The emigrant communities of Latvia. IMISCOE Research Series. Springer, Cham. https://doi.org/10.1007/978-3-030-12092-4_12

Kyaing KT (2012) Pull and push factors of migration: a case study in the urban area of Monywa Township, Myanmar. https://www.worldofstatistics.org/files/2014/03/Pull-and-Push-Factors-of-Migration-Thet.pdf

Kuschminder K (2017) Reintegration strategies: conceptualizing how return migrants reintegrate. Springer International Publishing

Matloob P (2015) Occupational choice of return migrants. IZA World of Labor 197:1–10. https://doi.org/10.15185/izawol.197

Mertin S, Herzberg DG, Weerasinghe S (2018) Global migration governance: existing architecture and recent developments. In: World migration report. IOM

MMC (2022) Why people migrate. Insights and key messages drawn from a decade of MMC research and 4Mi data collection. Mixed Migration Centre. http://www.mixedmigration.org/

Morawska E (2018) Qualitative migration research: viable goals, open-ended questions, and multidimensional answers. In: Zapata-Barrero R, Yalaz E (eds) Qualitative research in European migration studies. IMISCOE Research Series. Springer, Cham. https://doi.org/10.1007/978-3-319-76861-8_7

NIF (2019) Push or pull factors: what drives Central American migrants to the U.S.? National Immigration Forum. https://immigrationforum.org/wp-content/uploads/2019/07/PushPull-Factors_Paper_Final.pdf
Ndreka A (2019) Return migration and re-integration of returnees challenges in the origin country. RSC 11(3):4–24. https://doi.org/10.2478/rsc-2019-0012
Niedomysl T, Amcoff J (2010) Why return migrants return: survey evidence on motives for internal return migration in Sweden. Popul Space Place 17(5):656–673. https://doi.org/10.1002/psp.644
Nikolka T (2018) Family return migration. IFO Working Papers No. 286
OECD (2024) Return, reintegration and re-migration. OECD. https://doi.org/10.1787/625fb5e6-en
Pirvu R, Axinte G (2012) Return migration—Reasons, consequences and benefits, Annals of the University of Petroşani. Economics 12(4):193–202
Rašević M (2016) Migracije u funkciji razvoja: Mogućnost ili mit. In: Seobe i razvoj. Centar za ekonomska istraživanja Instituta društvenih nauka, pp 50–58
Rayan SI, Arcand J-L (2023) COVID-19 return migration phenomena: experiences from South and Southeast Asia. J Ethn Migr Stud 49:5133–5152. https://doi.org/10.1080/1369183X.2024.2268963
Sander M (2007) Return Migration and the "healthy immigrant effect". German Socio-Economic Panel Study (SOEP)
Šantić D, Todorović M, Perišić N (2021) The 'New Normal' in migration management in Serbia in times of the COVID-19 crisis. J Balkan near East Stud 24(3):1–19. https://doi.org/10.1080/19448953.2021.2015659
Simpson N (2017) Demographic and economic determinants of migration. IZA World of Labor 373. https://doi.org/10.15185/izawol.373
Stanković V (2014) Srbija u procesu spoljnih migracija. Republički zavod za statistiku RZS
Taran P, Tennant E, Simeone L (2017) Governance of migration. Global Migration Policy Associates GMPA
Urbański M (2022) Comparing push and pull factors affecting migration. Economies 10(1):21. https://doi.org/10.3390/economies10010021
Vanthomme K, Vandenheede H (2021) Factors associated with return migration of first-generation immigrants in Belgium (2001–2011). Eur J Popul 37(3):603–624. https://doi.org/10.1007/s10680-021-09581-5
Whyte Z, Hirslund DV (2013) International experiences with the sustainable assisted return of rejected asylum seekers. DIIS Report. DIIS, Copenhagen

Milica Langović is Research Associate at the Institute of Physics Belgrade, University of Belgrade, Serbia. She is particular interested in return migration and reintegration. She has been participating in a number of research projects on migration, integration, and governance research with a focus on creating policy recommendations. Langović published a number of academic articles and participated in numerous scientific conferences.

Danica Šantić is Full Professor at the Faculty of Geography, University of Belgrade, Serbia. Her academic focus is on migration issues. She has authored more than 70 scientific papers and participated in more than 20 research projects as well as attended numerous conferences. She is Co-founder of the Network for Migration, Interconnectivity and Regional Development MICaRD, Board Member of the Western Balkan Migration Network WB-MIGNET, Expert for the world bank, Research Fellow at the Canadian research chair for global migration processes, Country Member of the European Migration Network EMN Serbia, and World Social Science Fellow (WSSF) and Fellow of the European Centre for International Affairs ECIA. Prof. Šantić has worked as Expert for the Commission for Refugees and Migration CRM and many more.

Dejan Ž. Đorđević is Full Professor at the Faculty of Economics, University of Niš, Serbia. He has been working on more than ten international scientific projects. He is Author and Co-author of more than 70 scientific and research articles, one textbook, and one scientific monograph. His articles are published in national and international academic journals dedicated to the fields of Economic Geography, Tourism,and Demography.

Overcoming Inequality: Governance and Institutions

Chapter 6
Efficiency and Performance of Sustainable Development of the Western Balkan Countries: Whether Membership in the European Union is Recipe for Sustainability?

Predrag Mimović, Ana Krstić Srejović, and Milena Jakšić

Abstract To evaluate the impact of economic growth over time, the concept of sustainable development stresses the need to balance the three pillars of sustainability: economic, social, and environmental factors. This study compares the countries of the Western Balkans with former socialist countries now in the EU by analyzing their efficiency and overall performance in terms of sustainable development. The analysis focuses on data from 2019, just before the onset of the COVID-19 pandemic. The study utilized Data Envelopment Analysis (DEA) with a Slacks-Based Measure (SBM) model. The input variables in this model included carbon dioxide (CO_2) emissions, the unemployment rate, and energy consumption, while the output variables were Gross Domestic Product (GDP) and life expectancy at birth. The main finding of the study was the identification of the average efficiency scores, showing that EU member states typically had the highest scores and demonstrated the best practices, with Bulgaria and, to a lesser extent, Slovakia as notable exceptions, along with Albania. In contrast, most of the other countries in the Western Balkans were found to be inefficient in their sustainable development policies. Additionally, developing countries, on average, performed the worst across the variables analyzed. For a more detailed assessment of efficiency, the study also calculated the harmonized efficiency of the observed countries. This was done by dividing individual efficiency results by the average efficiency within the Western Balkan countries group and the

P. Mimović (✉) · M. Jakšić
Faculty of Economics, University of Kragujevac, Kragujevac, Serbia
e-mail: mimovicp@kg.ac.rs

M. Jakšić
e-mail: milenaj@kg.ac.rs

A. Krstić Srejović
Faculty of Economics, University of Belgrade, Belgrade, Serbia
e-mail: anakrstic@kg.ac.rs

J. J. Stanković et al. (eds.), *Regional and Urban Challenges and Opportunities for Data-Policy Interactions*, The Urban Book Series,
https://doi.org/10.1007/978-3-032-08592-4_6

EU member states group. Composite efficiency was also considered, which represents the weighted sum of DEA efficiency and inverse DEA efficiency. To determine the sustainable development performance of the countries, the simple additive weighting (SAW) method was used, applying the weighting coefficients derived from the CRITIC method. The results of this analysis were consistent with those obtained from the SBM DEA method. EU member states, except for Bulgaria and Slovakia, generally had the highest index values and were ranked the best in terms of sustainable development performance. On the other hand, the Western Balkan countries ranked significantly lower in comparison.

Keywords Sustainable development · Efficiency · Performance · SBM DEA · CRITIC · SAW · Ranking

6.1 Introduction

Achieving consensus regarding economic growth, environmental sustainability, and human well-being is considered one of the greatest challenges facing society today (Lira and Cândido 2008). As the global population continues to grow, the pressure on natural resources intensifies, and the complexities of maintaining a balance between economic development, environmental protection, and social equity become increasingly difficult. This tri-dimensional challenge underscores the need for a more holistic approach to development, one that not only focuses on economic prosperity but also ensures the preservation of our ecosystems and the improvement of social conditions for current and future generations. Sustainable development is a dynamic and evolving concept that requires a deep integration of economic growth, environmental sustainability, and social well-being. Achieving this balance is one of the most significant challenges facing humanity today, but it is also an opportunity to reimagine how societies function and how we measure success. By prioritizing ecological integrity, social equity, and long-term prosperity, sustainable development offers a path toward a more just and resilient world. Sustainability consists of three pillars: social, economic, and environmental (Camioto and Pulita 2022). These pillars form the "triple bottom line" (TBL), whose goal is to meet the resource needs of current and future generations without compromising the environment (Khan et al. 2021). That is, sustainable development is development that meets the needs of the present without compromising the ability of future generations to meet their own needs. Numerous studies (Santana et al. 2014; Camioto et al. 2016; Frugoli et al. 2015; Verma and Raghubanshi 2018; Zhou et al. 2019, etc.) have addressed the question of whether the observed growth model, environmental sustainability, and quality of life of the population are relevant variables. The concept of sustainable development, as a logical and consistent continuation of utopian theories of social well-being, social justice, and harmonious management of the environment and economic growth, establishes the closest relationship between socioeconomic development and sustainable development in its deepest form (Mukherjee and Chakraborti

2013; Kurniawan and Managi 2017). As things stand, the current model of economic growth completely collides with sustainable development, considering the literal depletion of many natural resources and the drastic degradation of environmental quality (Jorgenson and Dietz 2015). In this context, the concept of sustainable development is extremely important in developing countries to curb the unsustainable trend and direction of development in which modern society is moving. Therefore, it is crucial to analyze how these countries are developing to assess their performance in terms of sustainability and the quality of life for their populations.

The aim of this study is to compare the transition economies of the countries of the Western Balkans—Albania, Bosnia and Herzegovina, Montenegro, Northern Macedonia and Serbia—with the countries that were once part of the former Eastern bloc and are now members of the EU, i.e., Croatia, Hungary, Bulgaria, the Czech Republic, Slovakia, Romania, Slovenia, and Poland. The key result and contribution of the research are a ranking of the average efficiency of sustainable development, along with the averages of the relative disadvantages of each variable, showing the position of each country in relation to every variable used. These results allow an assessment of how each country performs in different areas of sustainable development, providing relevant information on which areas countries should focus on. For this purpose, the Data Envelopment Analysis (DEA) method was used, a non-parametric method for determining the so-called efficiency limits, based on linear programming. The DEA method will estimate the efficiency frontiers for a group of observed countries by analyzing multiple data sets and simultaneously solving multiple linear programming models. In line with the defined research focus and goals, this paper addresses the following research question:

RQ1: How does the average efficiency of sustainable development in the Western Balkan economies compare to that of former socialist states and current EU members in 2019?

RQ2: What is the overall sustainable development performance of the observed countries in 2019?

RQ3: What is the relationship between efficiency and overall sustainable development performance in the observed countries in 2019?

The paper is divided into four parts. The first part provides an introduction to the context of the research objective. The second part is a literature review related to the application of the *DEA* method to the topic of sustainable development. The third part briefly introduces the methods used in the study: *Slacks-Based Measure (SBM) DEA*, *Criteria Importance through Intercriteria Correlation* (*CRITIC*) and *simple additive weighting* (*SAW*) method. The fourth part refers to the clarification and structuring of the appropriate *SBM DEA* model to evaluate the efficiency of sustainable development and includes results and discussion that focus on the presentation of the results obtained in the study and their analysis.

Finally, to measure the performance of sustainable development as a complex socioeconomic phenomenon, a composite measure was calculated using the method of simple additive weights, using the values obtained by the method *CRITIC* as weighting coefficients of the criteria, and the results and ranks were compared with the results of the *DEA* model. Finally, there is a conclusion that concludes the study and shows how it can be used. Figure 6.1 shows all of the above.

6.2 Literature Review

The efficiency and effectiveness of sustainable development are frequently discussed in terms of observing, measuring, and analyzing the extent to which the Sustainable Development Goals (SDGs) are being achieved. These goals serve as a comprehensive framework for assessing global progress toward a more equitable, resilient, and environmentally sustainable future. The SDGs, consisting of 17 goals, were adopted by the United Nations in 2015 as part of the 2030 Agenda for Sustainable Development. This global initiative represents a call for international cooperation to eradicate poverty, safeguard the planet's ecosystems, and create the conditions for social well-being by 2030. These objectives encompass a wide range of issues, from economic development to environmental preservation and social equality. One of the key characteristics of the SDGs is their interconnected nature. Each goal is designed to support and enhance the others, recognizing that sustainable development cannot be achieved in isolation within one sector. For instance, efforts to combat climate change (Goal 13) are closely linked to the need for affordable and clean energy (Goal 7), which in turn impacts economic growth (Goal 8) and social inclusion (Goal 10). This interconnectedness highlights the complexity of measuring sustainable development as outcomes in one area inevitably influence others. As such, achieving balance across these domains is crucial, and the pursuit of progress in one goal must not come at the expense of another (Ganda and Ngwakwe 2014; Türkoglu and Kardogan 2017; Fankhauyer and Jotzo 2018; Zakari et al. 2022), ecological and eco-industrial efficiency (Kurniati et al. 2010; Unruh and Ettenson 2010; Figge et al. 2014; Caiado et al. 2017; Zielinska-Chmielewska et al. 2021; Long 2021; Wang et al. 2019; Sharma et al. 2020; Khan et al. 2021; Zhou et al. 2021; Liu et al. 2022; Figge and Thorpe 2023), innovation efficiency (Nidumolu et al. 2009; Carfì and Schilirò 2012a, b; Cheng et al. 2014; Luqmani et al. 2017; Schiliro 2019; Krstic and Mimovic 2018, 2020; Krstić et al. 2021; Zhang et al. 2022), educational efficiency (Lupi et al. 2022; Alméstar et al. 2022; Casciati et al. 2022; Mylonas et al. 2023), land-use efficiency (Wei et al. 2023; Song et al. 2023; Wang et al. 2023), etc. When evaluating the efficiency and effectiveness of sustainable development, it is essential to adopt a multidimensional approach that considers the broad scope of the SDGs. These aspects are typically analyzed through a combination of quantitative and qualitative indicators that measure progress, identify challenges, and highlight areas for improvement. Energy efficiency, for example, plays a central role in assessing the effectiveness of environmental goals, particularly given its direct impact on reducing greenhouse gas emissions, enhancing resource use, and supporting sustainable production and consumption patterns. Indicators related to energy efficiency can help assess how well countries are meeting their climate targets while also contributing to economic growth

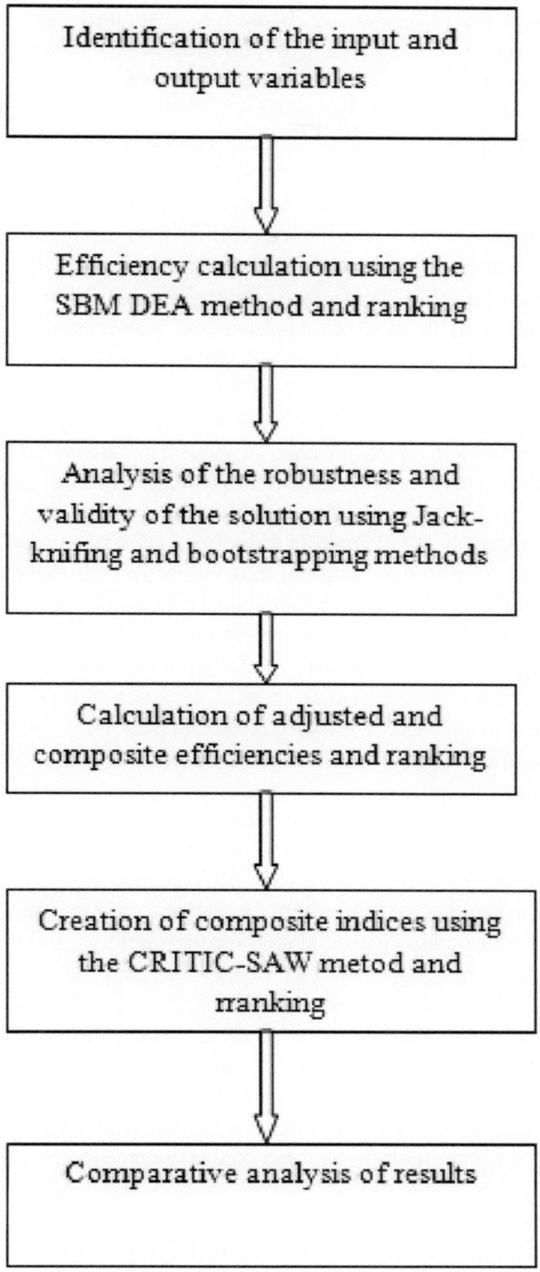

Fig. 6.1 Research framework of the work. *Source* Authors

and social well-being. Moreover, evaluating the efficiency of sustainable development involves looking beyond the achievement of individual goals to understand how they interact and contribute to overall sustainability. This involves considering trade-offs, synergies, and potential negative externalities that might arise when prioritizing certain goals over others. For example, while increasing industrial production may drive economic growth (Goal 8), it could also lead to greater environmental degradation if not paired with sustainable practices. Therefore, analyzing sustainable development performance requires a nuanced and integrated approach, ensuring that the strategies adopted are truly advancing long-term sustainability across all sectors. On the other hand, starting from the indicators of sustainable development, which are used to measure progress and achievements in various areas, and taking into account that these indicators integrate and balance three key dimensions of social development-economic, environmental, and social-a logical question arises as to whether it is possible to create a model that will adequately encompass all these dimensions in the process of measuring sustainable development performance. This question is important because sustainable development is not a one-dimensional concept, but rather encompasses a complex interaction between different aspects that must be balanced in order to achieve sustainability at the global level. Therefore, the need to find a model that will reflect all these aspects and measure them in the right way becomes crucial. In the literature dealing with this issue, a large number of studies focus on the selection and definition of appropriate indicators that will be useful for measuring sustainable development. These indicators should not only reflect the current state of affairs in each of the three dimensions, but also allow for the prediction of future trends, impacts and needs, which is necessary for planning and guiding policies that will contribute to long-term sustainability. The indicators must be selected in a way that will ensure that all critical areas of development are taken into account, from economic well-being, through environmental protection, to improving the quality of people's lives. Therefore, the process of structuring such a model requires the integration of different data and analyses that will be able to reflect the complexity and interdependence of these dimensions. This is a challenge, but at the same time a crucial step for the implementation of an effective system of assessment and monitoring of sustainable development (Santana et al. 2014; Rodriguez et al. 2018; Viglia et al. 2017; Biggeri and Mauro 2018; Mikayilov et al. 2018; Raza and Shah 2018; Cai et al. 2018; etc.). The number of indicators and their selection in the aforementioned studies varies depending on the objective and the chosen research method, and cannot be said to be predetermined. In this sense, the research results were not unambiguous and unconditionally generalized. A broader and good overview of studies in this context is provided by Camioto and Pulita (2022).

The DEA method has often been used in development performance analyses. A good overview of applications in this sense is given by Tsaples and Papathanasiou (2021) and Zhou et al. (2018). We will list some of them in order to illustrate the way and framework of using the DEA method and its extensions: Zhang et al. (2008) conducted an eco-efficiency analysis for regional industrial systems in China by developing Data Envelopment Analysis-based models. The results show that, provinces with higher level GDP per capita will have higher eco-efficiency relatively; Tian et al. (2019), use the SBM DEA model to measure the sustainable efficiency of regional transport through a system of indicators that includes systemic, economic, social and environmental indicators and obtain quite realistic results; Lee et al. use a two-stage DEA model to evaluate the efficiency of industrial waste recycling treatment, and indicate three efficient recycling models; Wang et al. (2018) on the example of the Chinese thermal power industry proposed a DEA in the evaluation of the effectiveness of reducing the emission of harmful gases; Zhou et al. (2021) used the SBM DEA model, Malmquist index model of productivity, and Tobit regression model to measure and evaluate the industrial eco-efficiency. The results show that it is positively correlated with industrial development level, scientific and technological innovations, industrial structure, and environmental planning, whereas negatively correlated with the degree of opening up and industrial agglomeration; Long (2021) evaluated the eco-efficiency of 35 major Chinese cities from 2011 to 2015 using a Slacks-Based Measured DEA model. The findings reveal that the overall efficiency of these cities during the observed period is relatively low, with improvements in total-factor ecological productivity largely driven by technological advancements; Zurano-Cervelló et al. (2019) combined life cycle assessment, Data Envelopment Analysis, and advanced mathematical programming to compare and optimize electricity mixes based on sustainability criteria, applied to the 28 EU member states. The findings suggest that EU energy policies should focus on expanding hydropower, wind, and solar energy while gradually eliminating non-renewable sources and bioenergy; Toma et al. (2017) assessed the agricultural efficiency of EU countries using a bootstrap-DEA approach. The study concludes that policy planning and management decisions should not only focus on maximizing agricultural output but also on preventing the overexploitation of environmental resources. Mimović et al. (2019) applied DEA to assess the relative efficiency of research and development in nine Southern European countries from 2007 to 2016. Their findings suggest that being a member of the European Union is not always a requirement for the efficient utilization of research and development resources and capacities at the national level; Krstic et al. (2022a) used a combined approach of DEA Window analysis and the Malmquist productivity index to measure the macroeconomic efficiency of Balkan countries from 2006 to 2018, using macroeconomic indicators as model variables. The results showed that during the observed period, Albania and, to a larger extent, Montenegro had the highest average efficiency, effectively using available resources to achieve GDP growth. On the other hand, EU member states, particularly Greece and Croatia, saw the highest overall productivity growth, which was largely attributed to improvements in technical efficiency.

6.3 Methodology

Data Envelopment Analysis (DEA) is a method used to assess the relative efficiency of decision-making units (DMUs) by calculating the ratio of the weighted sum of outputs to the weighted sum of inputs required to produce them (Mello and Gomes

2004). In the DEA method, efficiency values always range between 0 and 1, with a DMU considered efficient if its value reaches a reference limit, which is set at 1 to represent full efficiency (Charnes and Cooper 1985). The Charnes, Cooper, and Rhodes DEA model (CCR) estimates the efficiency of a DMUk by calculating the maximum value of the ratio between the weighted sum of output variables and the weighted sum of input variables. The weights reflect the relative importance of the input and output variables, and their optimal values are determined a posteriori, to ensure that the system of constraints is satisfied and that the objective function reaches its extreme value. The model applied in this paper is the *Slacks-Based Measured* (SBM) variant, which, as described by Tone (2001), seeks to minimize inputs while maximizing outputs simultaneously. This implies that reducing inputs does not necessarily lead to a decrease in outputs.

6.3.1 Slacks-Based Measure (SBM) DEA

Assume that there are n decision units, denoted as DMUs. Let DMU_j $(j = 1, 2,\ldots, n)$ consume m_I variable input $x_{i_{1j}}(i = 1, 2, .., m_1)$ and m_2 fixed input $x_{i_{2j}}(i = 1, 2, .., m_2)$, to produce k_I variable output $y_{r_{1j}}(r_1 = 1, 2, .., k_1)$ and k_2 fixed output $y_{r_{2j}}(r_2 = 1, 2, .., k_2)$. Then, according to Tone (2001), we can evaluate the performance of the observed DMU_j through the following SBM DEA model:

$$\min\theta = \frac{1 - \frac{1}{m_1}\sum_{i_1=1}^{m_1} \frac{s_{i_1}^-}{x_{i_1 d}}}{1 + \frac{1}{k_1}\sum_{r_1=1}^{k_1} \frac{s_{r_1}^+}{y_{r_1 d}}}. \tag{6.1}$$

Under restrictions:

$$\sum_{j=1}^{n} x_{i_1 j}\lambda_j + s_{i_1}^- = x_{i_1 d},\ i_1 = 1, \ldots, m_1, \tag{6.2}$$

$$\sum_{j=1}^{n} x_{i_2 j}\lambda_j = x_{i_2 d},\ \ i_2 = 1, \ldots, m_2 \tag{6.3}$$

$$\sum_{j=1}^{n} y_{r_1 j}\lambda_j - s_{r_1}^+ = y_{r_1 d},\ \ r_1 = 1, \ldots, k_1 \tag{6.4}$$

$$\sum_{j=1}^{n} y_{r_2 j}\lambda_j = y_{r_2 d},\ \ r = 1, \ldots, k_2 \tag{6.5}$$

$$x_{i_1 d} - s_{i_1}^- \geq L_{i_1} x_{i_1 d},\ \forall i_1, \tag{6.6}$$

$$y_{r_1d} + s_{r_1}^{+} \leq U_{r_1} y_{r_1d}, \forall r_1, \tag{6.7}$$

$$\lambda_j \geq 0, \forall j.$$

The optimal solution of model (6.1) is denoted as $\left(\theta^*, \lambda_j^*, s_{i_1}^{-*}, s_{r_1}^{+*}\right)$. Based on this solution, the performance of DMU_d can be improved by $x'_{i_{jd}} = x_{i_{jd}} - s_{i_1}^{-*}$ i $y'_{r_{jd}} = y_{r_{jd}} + s_{r_1}^{+*}$, wherein θ^* the efficiency of the d_{th} decision-making unit, λ_j^* contribution of the reference efficient decision-making unit in achieving the goal of the analyzed decision-making unit, and $\left(x'_{i_{jd}}, \forall i_1\right)$ i $\left(y'_{r_{jd}}, \forall r_1\right)$ optimally improved input and output variables, respectively. Constraints 2.5 and 2.6 limit the improved values of the variables within the corresponding range and affect the optimal solution. The model used seeks to minimize inputs while simultaneously maximizing outputs. This implies that reducing inputs does not necessarily result in a reduction of outputs.

6.3.2 *CRITIC Method*

The *CRITIC* method is a weighting approach used to determine objective weights for criteria in multi-criteria decision-making problems. Developed by Diakoulaki et al. (1995), this method takes into account contrast intensity and conflict within the decision problem's structure. *CRITIC* employs correlation analysis to identify contrasts between criteria (Yilmaz and Harmancıoğlu 2010). In this approach, the decision matrix is assessed, and the standard deviation of the normalized values for each criterion, along with the correlation coefficients between all pairs of criteria, is used to determine the contrast of each criterion, C_j. The method assigns greater weight to criteria that have a high standard deviation and low correlation with other criteria. Essentially, a higher value of C_j indicates that more information is derived from that criterion, making it more important for the decision problem.

6.3.3 *Simple Additive Weighting*

Simple additive weighting (*SAW*) is a method used in multi-attribute decision-making to select the most favorable alternative. The main idea behind the SAW method is to calculate the maximum sum of the weighted performance scores for each alternative across all attributes (Khairul and Siapun 2016). This method typically involves linearly normalizing the decision matrix, and the weight coefficients can be determined either through subjective or objective methods.

6.4 Structuring the Model and Results

In this study, the DEA method will be applied to analyze the efficiency of the observed countries in implementing sustainable development policies, focusing on achieving an optimal balance between economic growth, social well-being, and minimizing the negative environmental impact. The variables used in this research were selected for this purpose, particularly because they align with the three pillars of sustainable development, as indicated in the relevant studies, and due to the availability of data for all countries. The analysis covers the year 2019, which was just before the COVID-19 pandemic and the conflict in Ukraine, events that significantly impacted economic trends and, by extension, other areas of life such as energy consumption and employment. The assumptions underlying the DEA model are as follows: (1) The analysis covers the year 2019; (2) the input variables include: CO_2 emissions (in millions of tons per capita), energy consumption (in kilograms of oil equivalent per 1000 Euros of GDP), and unemployment rate (%). The output variables are: GDP growth rate (%), which is the most commonly used indicator of a country's overall economic activity, and life expectancy at birth (Table 6.1). Data for the selected input and output indicators were collected from the *World Bank data* and EUROSTAT databases.

In line with the principles of the SBM DEA method, the goal of our model is to reduce CO_2 emissions, unemployment rate, and energy consumption, while aiming to increase GDP and life expectancy at birth. DEA Solver was utilized to conduct this analysis (Table 6.2).

The results obtained by applying the SBM DEA model (efficiency of DMU_s and slacks of the variables), are shown in Table 6.5. It has been shown that the member countries of the European Union (Croatia, Hungary, Romania, Slovenia, and the Czech Republic) are efficient in terms of sustainable development and ranked first, with the exception of Bulgaria and Slovakia, which are 69% and 79% of the reference efficiency, respectively. The surprise is Albania, which also ranks at the level of best practice in terms of efficiency and is ranked first. Along with Poland, it is one of the only countries with efficiency above the average for the entire EU sample (0.92). For a more complete and comprehensive assessment of efficiency, the harmonized efficiency of the observed countries was calculated, by comparing the individual efficiency values of each country with the average efficiency within two different groups: the Western Balkans (0.64) and the EU Member States (0.92). This approach allows for a deeper understanding of how countries perform relative to each other within these regional categories. In addition, to explore the impact of external factors on the efficiency of sustainable development, an adjusted efficiency score was calculated. This adjustment was made by taking the ratio of each country's efficiency score to the average efficiency score of the corresponding groups (Western Balkans at 0.64 and EU Member States at 0.92). As illustrated in Table 6.11, after this adjustment, the efficiency of EU member states, with their efficiency measure value of 1, showed little or no change, highlighting that their performance was already close to the average for their group. In contrast, the efficiency scores for non-EU countries, especially in the Western Balkans, experienced a significant increase, with Albania

Table 6.1 Input and output variable for DEA

	Variable	Explanation	Source of data
Input variables	CO_2 emissions	The carbon dioxide produced during the consumption of solid, liquid, and gaseous fuels, as well as the combustion of gas	https://data.worldbank.org/
	Energy consumption	The use of primary energy before it is converted into other fuels for final consumption	https://data.worldbank.org/ https://ec.europa.eu/eurostat
	Unemployment rate	The portion of the labor force that is unemployed but available to seek work	https://data.worldbank.org/
Output variables	GDP %	The sum of the gross value added of all resident producers in the economy, increased by taxes on products and decreased by all subsidies not included in the value of the product	https://data.worldbank.org/
	Life expectancy	Years of life if there were a trend of current mortality patterns throughout life	https://ec.europa.eu/eurostat

Source Authors

Table 6.2 Statistics on input/output data

Variable	CO_2 emissions	Energy consumption	Unemployment rate	GDP %	Life expectancy
Max	9	458	17.3	4.9	82
Min	1.7	169.6	2	2.1	75
Average	5.3846	280	7.9692	3.6	77.538
S.D.	1.8425	98.624	5.1474	0.7805	1.7371

Source Authors

recording the most significant improvement its efficiency score rising from 1 to 1.56. This adjustment process revealed that the top-ranking countries generally showed both strong operational efficiency and high relative efficiency, especially those that are part of the EU, indicating a correlation between EU membership and better performance in sustainable development. On the other hand, some EU member states, such as Bulgaria and Slovakia, were inefficient, showing low operational efficiency and weak relative efficiency. This suggests that EU membership does not necessarily guarantee optimal performance in terms of sustainable development. Interestingly, Albania, a non-EU country, emerged as an exception to the pattern. Despite not being an EU member, Albania showed a relatively high efficiency score, especially compared to its Western Balkan counterparts. This anomaly highlights the complexity of efficiency dynamics and suggests that factors outside EU membership, such as the quality of governance or strategic national policies, can significantly influence a country's efficiency in sustainable development. This adjusted efficiency analysis not only highlights the different levels of performance between countries within these two groups, but also highlights the importance of considering both operational and relative efficiency when assessing sustainable development outcomes. The results suggest that regional context, as well as external factors, plays a key role in shaping countries' efficiency in the domain of sustainable development. Moreover, looking at Albania's performance challenges the assumption that EU membership is a definitive factor in achieving high efficiency, pointing to the need for a more nuanced understanding of what drives success in sustainable development in different regions. Changes on the side of the output variables are limited to the GDP growth rate, namely in Bosnia and Herzegovina where in 2019 a growth rate of 3.95% was necessary, instead of the achieved 2.8%, and in Slovakia, 4% instead of the achieved 2.5%. At the same time, in Bosnia and Herzegovina it was necessary to reduce CO_2 emissions by 38.89%, energy consumption by 54.56%, and unemployment by 74.8%, in order for sustainable development to be at the level of the best practices of efficient countries. Slovakia needed to reduce CO_2 emissions by 31.57%, energy consumption by 0.187% and unemployment by 30.9% in order to achieve effective sustainable development in 2019, while, for example, Serbia had to emit 36.97% less carbon dioxide, to reduce energy consumption by 52.50% and unemployment from 10.4 to 3.7%, which is a reduction of 64.4%, so that in 2019 the efficiency of its sustainable development would be at the level of best practice. In the other observed countries, the necessary changes were exclusively on the side of input variables. On average, at the level of the entire group of observed countries, it was necessary to reduce CO_2 emissions by 13.75%, energy consumption by 17.33%, unemployment by 25.44% and at the same time increase the average GDP growth rate from 3.53 to 3.8%, so that the efficiency of sustainable development would be equal1. It is clear that, to the greatest extent, the necessary changes in input variables were limited to non-EU countries, along with Bulgaria and Slovakia (Table 6.3). Average values original and optimal value of input and output variables are shown in Table 6.4.

It is interesting that the comparative analysis of the efficiency of sustainable development in 2019 and 2021 showed that the deviations in the values and ranking of countries are insignificant (Fig. 6.2). In doing so, it was assumed that life expectancy in a short period remained at the level of 2019, while some values of the selected variables, primarily the countries of the Western Balkans, for which data in 2021 are not available, were corrected for the average change at the global level compared to the previous year. For example, it is known that the average level of energy consumption growth in 2021 at the global level was higher by about 5% compared to 2020, in which the level was lower by about 4% compared to 2019.With the conditional acceptance of the aforementioned assumptions, the analysis showed that in 2021, the EU

Table 6.3 Original and optimal value of input and output variables

	CO_2 emissions			Energy consumption			Unemployment rate			GDP %			Life expectancy		
DMU	Data	Project	Diff%	Data	Project	Diff %	Data	Project	Diff%	Data	Project	Diff%	Data	Project	Diff %
Albania	1.7	1.7	0	224.6	224.6	0	11.5	11.5	0	2.1	2.1	0	79	79	0
Bosnia and H	6.3	3.85	−38.88	458	208.1	−54.56	15.7	3.95	−74.8	2.8	3.95	41.1	77	77	0
Bulgaria	5.6	3.88	−30.60	424.4	206.4	−51.35	4.2	3.77	−10.1	4	4	0	75	75	0
Croatia	4.1	4.1	0	185.9	185.9	0	6.6	6.6	0	3.4	3.4	0	78	78	0
Hungary	4.7	4.7	0	230.1	230.1	0	3.4	3.4	0	4.9	4.9	0	76	76	0
Montenegro	4.2	3.98	−5.1	281.5	211.7	−24.76	15.1	3.87	−74.3	4.1	4.1	0	77	77	0
Romania	3.8	3.8	0	205.4	205.4	0	3.9	3.9	0	3.9	3.9	0	76	76	0
N.Mac	4	3.79	−5.2	334	208.3	−37.60	17.3	4.14	−76.0	3.9	3.9	0	77	77	0
Slovenia	6.5	6.5	0	169.6	169.6	0	4.4	4.4	0	3.5	3.5	0	82	82	0
Serbia	6.6	4.16	−36.97	453.3	215.2	−52.50	10.4	3.7	−64.4	4.3	4.3	0	76	76	0
Czech	9	9	0	238.5	238.5	0	2	2	0	3	3	0	79	79	0
Slovakia	5.7	3.9	−31.6	211.2	210.8	−0.187	5.8	4.00	−30.9	2.5	4.0	60.1	78	78	0
Poland	7.8	5.42	−30.39	231.6	230.5	−0.439	3.3	3.3	0	4.4	4.4	0	78	78	0

Source Authors

Table 6.4 Average values original and optimal value of input and output variables

	CO_2 emissions			Energy consumption			Unemployment rate			GDP %			Life expectancy		
	Data	Project	Diff%	Data	Project	Diff %	Data	Project	Diff%	Data	Project	Diff%	Data	Project	Diff%
Aver.	5.38	4.5	−13.75	280.6	211.1	−17.03	7.96	4.50	−25.44	3.6	3.8	7.78	77.5	77.5	0
Max	9	9	0	458	238.5	0	17.3	11.5	0	4.9	4.9	60.1	82	82	0
Min	1.7	1.7	−38.88	169.6	169.6	−54.56	2	2	−76.03	2.1	2.1	0	75	75	0
St dev	1.91	1.7	16.66	102.6	18.5	23.50	5.35	2.32	33.79	0.8	0.6	19.4	1.80	1.80	0

Source Authors

Table 6.5 Efficiency and slack variables

			Slack	Slack	Slack	Slack	Slack
DMU	Eff	Rank	CO_2 emissions	Energy consumption	Unemployment rate	GDP %	Life expectancy
Albania	1	1	0	0	0	0	0
Bosnia and H	0.439	13	2.45	249.897	11.749	1.151	0
Bulgaria	0.692	9	1.714	217.965	0.427	0	0
Croatia	1	1	0	0	0	0	0
Hungary	1	1	0	0	0	0	0
Montenegro	0.652	10	0.216	69.725	11.223	0	0
Romania	1	1	0	0	0	0	0
N.Mac	0.603	11	0.209	125.606	13.153	0	0
Slovenia	1	1	0	0	0	0	0
Serbia	0.487	12	2.44	238.02	6.7	0	0
Czech	1	1	0	0	0	0	0
Slovakia	0.790	8	1.8	0.395	1.797	1.503	0
Poland	0.897	7	2.371	1.017	0	0	0

Source Authors

member states were at the level of the efficiency limit in terms of sustainable development and, with the exception of the Slovak Republic, maintained their rank, while the countries of the Western Balkans were more or less also at their level of efficiency immediately before the corona virus pandemic. Albania also showed the resilience of its economy, and thanks to the relatively proportional, mutually balancing changes in the values of the selected variables, in the period 2019–2021, it remained at the level of the efficiency limit, maintaining a high rank of 1. It should also be noted that the overall average efficiency is at the level of the observed group, in 2021 amounted to 87.6% of reference efficiency, which is above 81.2% in 2019, which confirms the thesis about the sustainability of the observed economies despite the initial shock caused by the corona virus pandemic. However, it seems that according to the available data on the movement of the values of the selected variables in 2022, it seems that the situation was not rosy (Tables 6.5 and 6.6).

Analysis of the Robustness and Validity of the Solution

The reliability and robustness of the position of the best-performing units can be tested in several ways. One of them is *Jack-knifing* analysis. In this analysis, one efficient DMU is alternately eliminated from the set of DMUs in order to establish a possible difference in efficiencies (Charles et al. 2012). From the results obtained by applying *Jack-knifing* analysis, by alternating elimination of efficient DMUs, it can be concluded that there are no significant changes in the position of efficient DMUs, and that they remain effective despite the mentioned changes in the set

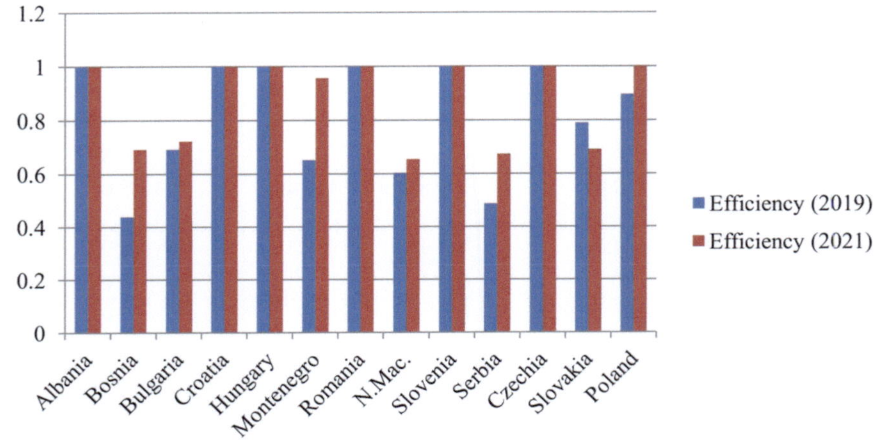

Fig. 6.2 Comparative presentation of efficiency of sustainable development of the observed countries in 2019 and 2021. *Source* Authors

Table 6.6 Average values of slack variables

	Score	Rank	CO_2 emissions	Energy consumption	Unemployment rate	GDP %	Life expectancy
Average	0.8126	5.8462	0.8615	69.4327	3.4653	0.2042	0
Max	1	13	2.45	249.897	13.153	1.503	0
Min	0.4391	1	0	0	0	0	0
St Dev	0.2136	4.9134	1.0877	101.89	5.2376	0.5035	0

Source Authors

of DMUs, which confirms the robustness of the original results obtained. One of the challenges associated with DEA is that the efficiency estimates obtained can be sensitive to the sample size and the specific data points used. This sensitivity can lead to biased estimates, especially in cases where the sample is small or not fully representative of the wider population. Biases in DEA efficiency results can arise from various factors, such as the influence of outliers, data dimensionality, and measurement errors, which can distort the true efficiency levels of the DMUs being estimated. To address these issues and improve the accuracy of efficiency estimates, the Bootstrap method can be used, which involves resampling the original data with replacement to generate multiple new samples (called "bootstrap samples") that mimic the characteristics of the original data set. This process allows for the creation of a sampling distribution for the DEA efficiency results, which can then be used to assess the variability and bias of the original estimates. These adjusted estimates take into account sampling variability and provide a more robust estimate of the performance of the DMU. In addition to reducing bias, the bootstrap method also improves the interpretability of the DEA results. Although DEA can provide point estimates of efficiency, these estimates alone may not convey the full uncertainty associated with the evaluation. The bootstrap approach allows for the calculation of confidence intervals, which provide a range of possible values for efficiency rather than a single point estimate. This additional layer of information allows decision makers to assess the reliability of the DEA results and make better-informed decisions based on the degree of uncertainty inherent in the data (Krstić et al. 2021, 2022b; Rejman-Petrović et al. 2022; Jakšić et al. 2023, etc.).

The bias value is calculated as the difference between the bootstrap value and the calculated parameter value, i.e.,

$$\mathrm{Bias}(\theta_k) = E(\theta_k) - \theta_k,$$

where $E(\theta_k)$ represents the bootstrap value of the parameter, in this case, the average efficiency, and θ_k is the measured average efficiency. The model efficiency, corrected for bias estimation, is calculated as the difference[1]:

$$\theta_k^{\mathrm{corr}} = 2 \times \theta_k - E(\theta_k).$$

For $n = 1000$ pseudo samples, with a confidence level of 0.05 the corrected average efficiency was calculated:

$$\theta_k^{\mathrm{corr}} = 2 \times 0.8126 - 0.796 = 0.829$$

which is slightly higher than the calculated average DEA efficiency, which also confirms the validity and accuracy of the obtained DEA results.

Composite Efficiency and Overall Performance of Sustainable Development

To allow for greater discrimination in terms of efficiency across countries or decision-making units (DMUs), the so-called inverse efficiency frontier method is often used (Entani et al. 2002; Angulo-Meza et al. 2005; Soares de Mello et al. 2008; Camioto and Pulita 2022). This method provides a more nuanced and detailed analysis of efficiency by addressing one of the inherent limitations of traditional DEA models, which tend to group DMUs with different levels of inefficiency into a single efficiency frontier. While conventional DEA focuses on constructing a frontier based on the best-performing units, the inverse efficiency frontier method seeks to identify inefficiency by constructing a frontier composed of the units that exhibit the worst management practices, often referred to as the inefficiency frontier. Essentially, the inverse efficiency frontier method works by reversing the standard DEA approach. Instead of focusing primarily on the best-performing units to establish efficiency measures, it identifies those units with the least efficient management practices. These inefficient units are used to construct a frontier that represents the worst-case performance scenario. The inefficiency frontier provides a reference point that allows for a clearer distinction between more efficient and less efficient units within the sample. In

[1] Song et al. (2013, p. 52).

addition, the inverse efficiency frontier method is particularly useful when the focus is on identifying "best practice" inefficiencies, which can inform policy decisions and guide improvements in management strategies. By analyzing the inefficiency frontier, policymakers and practitioners can identify key factors contributing to poor performance and target areas for intervention. This allows for more focused and effective strategies aimed at reducing inefficiencies and improving overall performance in specific sectors or countries. The inverse efficiency frontier method is also useful for distinguishing between countries or organizations that, while performing adequately, may still be far from optimal in terms of resource allocation, production processes, or management practices. This helps to uncover inefficiencies that might otherwise remain hidden when more traditional methods are used, leading to more accurate estimates of performance across a range of dimensions. To calculate the inefficiency limit, the inputs are replaced with the outputs of the original DEA model, and the inverse efficiency values are shown in Table 6.7.

With the results of the inverse limit, a composite index (*CI*) was calculated, as:

$$CI = \alpha \times \text{Eff} + (1 - \alpha) \times (1 - \text{InvEff}),$$

where α is the confidence index, whereby for different values of α, different values of the composite index of efficiency can be obtained. So, for the values α = (0.3, 0.5, and 0.7), respectively, the value of the composite index is shown in Table 6.8.

For $\alpha = 0.5$, the average value of the composite index of all observed countries is 0.47, while the average CI value of the countries of the Western Balkans and EU members is 0.35 and 0.55, respectively. As expected, the EU member states have the highest *CI* value, with the exception of Slovakia and Bulgaria, whose CI values are

Table 6.7 Inverse efficiency

DMU	Score	Rank
Albania	0.845	5
Bosnia and Herzegovina	1	1
Bulgaria	0.8	7
Croatia	0.528	11
Hungary	0.509	12
Montenegro	0.809	6
Romania	0.465	13
North Macedonia	1	1
Slovenia	0.705	10
Serbia	1	1
Czech	1	1
Slovakia	0.789	8
Poland	0.758	9

Source Authors

Table 6.8 Composite efficiency of observed countries

DMU	CI^a	CI^b	CI^c
Albania	0.409	0.578	0.747
Bosnia and Herzegovina	0.130	0.220	0.307
Bulgaria	0.348	0.445	0.544
Croatia	0.630	0.736	0.842
Hungary	0.640	0.746	0.847
Montenegro	0.329	0.421	0.513
Romania	0.670	0.500	0.860
North Macedonia	0.181	0.300	0.422
Slovenia	0.794	0.650	0.788
Serbia	0.146	0.244	0.341
Czech Republic	0.300	0.500	0.700
Slovak Republic	0.385	0.246	0.616
Poland	0.438	0.570	0.700

Note [a] $-(\alpha = 0.3)$; [b] $-(\alpha = 0.5)$; [c] $-(\alpha = 0.7)$
Source Authors

below the average not only for the group of EU member states, but also below the average of all countries. Also, with the countries of the Western Balkans, there are no unexpected results, except for Albania, whose *CI* value is even above the average at the level of the observed EU member states (Table 6.9).

The overall performance (*OP*) of sustainable development in the observed countries was determined using the simple additive weighting method, with the weights for the criteria derived from the *CRITIC* method. When assigning weights to the criteria, both the standard deviation and the correlation between the criteria are taken into account. In this context, the weight of the jth criterion (w_j) is calculated as follows:

$$w_j = \frac{C_j}{\sum_{j=1}^{n} C_j}, \tag{6.8}$$

Table 6.9 Weight coefficients of sustainable development factors calculated using the *CRITIC* method

Variable	w_j
CO_2 emissions	0.190
Energy consumption	0.182
Unemployment rate	0.213
GDP	0.238
Life expectancy	0.176

Source Authors

where C_j is the amount of information contained in the jth criterion and is calculated as:

$$C_j = \sigma_j \sum_{j'}^{n} (1 - r_{jj'}), \tag{6.9}$$

where σ_j is the standard deviation of the jth criterion and r_{jj} is the correlation coefficient between the two criteria, the normalized decision matrix. The decision matrix is normalized using the following equation:

$$x_{ij}^* = \frac{x_{ij} - \min(x_{ij})}{\max(x_{ij}) - \min(x_{ij})} \quad i = 1, 2, .., m; \; \mathrm{j} = 1, 2, .., n \tag{6.10}$$

for the benefit criteria, and

$$x_{ij}^* = \frac{\max(x_{ij}) - x_{ij}}{\max(x_{ij}) - \min(x_{ij})} \quad i = 1, 2, .., m; \; \mathrm{j} = 1, 2, .., n \tag{6.11}$$

for the non-benefit criteria,

wherein x_{ij}^* is the normalized value of the performance of the alternative i on the criterion j, and x_{ij} is the value of the performance of the alternative i on the criterion j, where it should be noted that the normalization does not take into account the type of criterion.

The weight coefficients of the selected variables, calculated using the CRITIC method, are shown in Table 6.9.

In that case, the sustainable development performance of the observed countries, using the *CRITIC-SAW* method, is calculated as follows:

$$I_i^* = \sum_{j=1}^{5} x_{ij} w_j \; i = 1, \ldots, 13,$$

where

I_i^* is the sustainable development index, x_{ij} is the realized and normalized value of the ith country according to the jth dimension of sustainable development expressed through the corresponding variable, and w_j is the weighting coefficient of the j variable calculated using the CRITIC method. The values of the overall performance of sustainable development are shown in Table 6.11. It can be seen that the best performance, as expected, is achieved by the member states of the European Union, Slovenia (0.722), Hungary (0.712), Romania (0.659), Poland (0.638), Croatia (0.634), and Czech Republic (0.528). Those same countries, with slight deviations in the ranking, also have the highest efficiency in terms of sustainable development. It is also striking that Bulgaria, as a member of the EU, ranks quite poorly according to the realized values of the index of efficiency and performance of sustainable development, in ninth and tenth places, while Slovakia is in eighth place according to all observed parameters. As expected, the countries of the Western Balkans are ranked significantly worse according to the achieved composite indicators of sustainable development, with the exception of Albania, whose efficiency of sustainable development is equal to the best practice (1), while according to the sustainable development index, it is in a high seventh place. Albania's high position can be explained by relatively low values of input variables and relatively high values of output variables of sustainable development (Table 6.10). Mann–Whitney U-test of statistical significance of the difference between the SBM DEA efficiency values and the matched efficiency values shows that the z-score is -2.0708 and the p-value is 0.03837, which means that the result is significant at $p < 0.05$, that is H_0 (there is difference) is rejected. A similar conclusion (the z-score is 3.1992 and the p-value is 0.001378) is also valid in the case of testing the statistical significance of the differences in SBM DEA efficiency and composite efficiency, which indicates a high influence of the observed countries' membership in the European Union on their efficiency in terms of sustainable development. On the other hand, correlation analysis (H_0 5%) shows a high degree of correlation between the performance of sustainable development and the sustainable development of the observed countries in terms of efficiency, except with adjusted efficiency (Table 6.11).

Table 6.10 Comparative overview of efficiency, composite efficiency, and sustainable development performance of observed countries (2019)

DMU	Efficiency	Rank	Adjusted efficiency	Rank	*CI* (α = 0.5)	Rank	Sustainable Development Index	Rank
Albania	1	1	1.56	1	0.578	4↓	0.518	7↓
Bosnia and Herzegovina	0.439	13	0.69	13	0.220	13	0.176	13
Bulgaria	0.693	9	0.75	12↓	0.445	8↑	0.452	10↓
Croatia	1	1	1.09	2↓	0.736	2	0.634	5↓
Hungary	1	1	1.09	2↓	0.746	1↑	0.712	2↓
Montenegro	0.653	10	1.02	7↑	0.421	9↓	0.487	9
Romania	1	1	1.09	2↓	0.500	6↓	0.659	3↑
North Macedonia	0.604	11	0.94	9↑	0.300	10↓	0.411	11↓
Slovenia	1	1	1.09	2↓	0.650	3↓	0.722	1↑
Serbia	0.487	12	0.76	11↑	0.244	12↓	0.392	12
Czech Republic	1	1	1.09	2↓	0.500	6↓	0.528	6
Slovak Republic	0.791	8	0.86	10↓	0.246	11↓	0.512	8↑
Poland	0.897	7	0.98	8↓	0.570	5↑	0.638	4↑

Source Authors

Table 6.11 Means, standard deviations, and correlation coefficients of sustainable development performance and sustainable development efficiency (2019)

	Mean	S.D	1	2	3	4
SBM DEA efficiency	0.8126	0.2135	–	0.743	0.829	0.868
Adjusted efficiency	1.0008	0.2225		–	0.628	0.577
Composite efficiency	0.4735	0.1822			–	0.880
Sustainable Development Performance	0.5262	0.1516				–

Note $p < 0.05$
Source Authors

6.5 Conclusion

The necessity to align social development policymakers with the three pillars of sustainability (economic, social, and environmental) arises from the goal of achieving economic growth without causing environmental harm and ensuring an optimal level of well-being for the population. A particular challenge for underdeveloped and developing countries is to achieve these goals simultaneously. Therefore, the main objective of the work was to compare, within the framework of sustainable development, the more developed countries, the former members of the so-called Eastern Bloc, and current members of the EU, with the countries of the Western Balkans, in order to identify and analyze the comparative advantages of membership in the EU for sustainable development policies. Analyzing the achieved efficiency of sustainable development in 2019, it can be noted that the top five countries (Croatia, Hungary, Romania, Slovenia, and the Czech Republic) are part of the European Union both in terms of efficiency and composite efficiency, as well as in terms of sustainable development performance. However, it can also be noted that the remaining two EU members, Bulgaria and Slovakia, are ineffective in both directions studied, which shows that these countries need to improve their practices by investing more in sustainable development policies. On the other hand, among the Western Balkan countries that are not members of the EU, Albania stands out, having effectively implemented sustainability policies in 2019 in the context of the analyzed variables, with a relatively good sustainable development performance, while among the remaining countries, Serbia is an unpleasant surprise, occupying a convincing penultimate place according to all observed parameters. In the partial analysis of the achieved values of the variables in 2019, one can see the good results of the EU member states, with the exception of Bulgaria, and the poor results of the countries of the Western Balkans, with the exception of Albania. As for CO_2 emissions, the highest CO_2 emissions in 2019 were achieved by the Czech Republic and the lowest by Albania. Bosnia and Herzegovina had the highest energy consumption in 2019, while Slovenia had the lowest. Northern Macedonia had the highest unemployment rate in 2019, while the Czech Republic had the lowest. Hungary had the highest growth rate in 2019, while Albania had the lowest. Finally, Slovenia had the longest life expectancy at birth in 2019 and Bulgaria the shortest. Thus, it can be seen that at least one EU Member State ranked first or second in all variables. With the exception of Albania, all the countries of the Western Balkans, assuming effective sustainable development policies in 2019, the time frame of the analysis, showed a relatively significant decrease in the value of the input variables in the first place, while relatively good results were obtained in the values of the output variables. Among the EU member states, Bulgaria showed by far the worst performance in the achieved values of the input variables, which affected not only efficiency but also the performance of sustainable development. Therefore, if we compare the countries of the Western Balkans with the former members of the Eastern Bloc that are now members of the EU, we can see that the latter, with the exception of Bulgaria, were outstanding in terms of sustainability in 2019, which shows that they are skillful in taking advantage of EU membership and have policies that reveal a strong social awareness of sustainable development. It should also be said that the synergistic effect of simultaneous changes in the values of the observed variables of sustainable development in the period 2019–2021 was such that the observed countries coped relatively well and quickly recovered from the consequences of the corona virus pandemic, maintaining a high level of efficiency of sustainable development in a relative sense, while for some deeper and more specific conclusions, an analysis over a longer time horizon is necessary, as well as a comparison with developed countries, primarily of the European Union. In summary, the efficiency and effectiveness of sustainable development cannot be fully captured through isolated measures. Instead, they must be viewed through a comprehensive lens that takes into account the multidimensional nature of the SDGs and the ways in which progress in one area can either support or undermine objectives in others. By using a combination of robust indicators, including those for energy efficiency, we can better understand the degree of progress toward a more sustainable and equitable global society.

References

Alméstar M, Sastre-Merino S, Velón P, Martínez-Núñez M, Marchamalo M, Calderón-Guerrero C (2022) Schools as levers of change in urban transformation: practical strategies to promote the sustainability of climate action educational programs. Sustain Cities Soc 87:104239. https://doi.org/10.1016/j.scs.2022.104239

Angulo-Meza L, Biondi L, No Mello JC, Gomes EG (2005) ISYDS—Integrated system for decision support (SIAD Sistema Integrado de Apoio a Decisão): a software package for data envelopment analysis model. Pesquisa Operacional 25(3):493–503

Biggeri M, Mauro V (2018) Towards a more 'Sustainable' human development index: integrating the environment and freedom. Ecol Ind 91:220–231

Cai Y, Sam CY, Chang T (2018) Nexus between clean energy consumption, economic growth and CO_2 emissions. J Clean Prod 182:1001–1011

Caiado RGG, de Freitas Dias R, Mattos LV, Quelhas OLG, Filho W (2017) Towards sustainable development through the perspective of eco-efficiency—A systematic literature review. J Clean Prod 165:890–904

Camioto FC, Rebelatto DAN, Rocha RT (2016) Análise da eficiênciaenergéticanospaíses do BRICS: um estudoenvolvendo a AnáliseporEnvoltória de Dados. Gestão & Produção 23(1):192–203

Camioto FC, Pulita AC (2022) Efficiency evaluation of sustainable development in BRICS and G7 countries: a data envelopment analysis approach. Gestão & Produção 29:e022. https://doi.org/10.1590/1806-9649-2022v29e022

Carfì D, Schilirò D (2012a) A coopetitive model for the green economy. Econ Modell 29(4):1215–1219. https://doi.org/10.1016/j.econmod.2012.04.005

Carfì D, Schilirò D (2012b) Global Green economy and environmental sustainability: a coopetitive model. Commun Comput Inf Sci 300:593–606. Springer, Berlin

Casciati F, Casciati S, Faravelli L (2022) Resilience and sustainability for educational buildings. J Infrastruct Intell Resil 1(1):100005. https://doi.org/10.1016/j.iintel.2022.100005

Charles V, Kumar M, Kavitha SI (2012) Measuring the efficiency of assembled printed circuit boards with undesirable outputs using data envelopment analysis. Int J Prod Econ 136(1):194–206. https://doi.org/10.1016/j.ijpe.2011.11.010

Charnes A, Cooper WW (1985) Preface to topics in data envelopment analysis. Ann Oper Res 2:59–94

Cheng CC, Yang CL, Sheu C (2014) The link between eco-innovation and business performance: a Taiwanese industry context. J Clean Prod 64:81–90. https://doi.org/10.1016/j.jclepro.2013.09.050

Diakoulaki D, Mavrotas G, Papayannakis L (1995) Determining objective weights in multiple criteria problems: the CRITIC method. Comput Oper Res 22(7):763–770. https://doi.org/10.1016/0305-0548(94)00059-H

Entani T, Maeda Y, Tanaka H (2002) Dual models of interval DEA and its extensions to interval data. Eur J Oper Res 136:32–45

Fankhauser S, Jotzo F (2018) Economic growth and development with low-carbon energy. Wiley Interdiscip Rev: Clim Change 9(1):1–16. https://doi.org/10.1002/wcc.495

Figge F, Young W, Barkemeyer R (2014) Sufficiency or efficiency to achieve lower resource consumption and emissions? The role of the rebound effect. J Clean Prod 69(15):216–224. https://doi.org/10.1016/j.jclepro.2014.01.031

Figge F, Thorpe A (2023) Circular economy, operational eco-efficiency, and sufficiency. An integrated view. Ecol Econ 204(Part B):1–11, 107692. https://doi.org/10.1016/j.ecolecon.2022.107692

Frugoli PA, Almeida CMVB, Agostinho F, Giannetti BF, Huisingh D (2015) Can measures of well-being and progress help societies to achieve sustainable development? J Clean Prod 90:370–380

Ganda F, Ngwakwe CC (2014) Role of energy efficiency on sustainable development. Environ Econ 5(1):86–99

Jakšić M, Srejović AK, Milanović M, Mimović P (2023) Measuring the economic performance of transition economies: DEA-bootstrapping approach. J Econ Stud ahead-of-print No:ahead-of-print. https://doi.org/10.1108/JES-01-2023-0037

Jorgenson AK, Dietz T (2015) Economic growth does not reduce the ecological intensity of human well-being. Sustain Sci 10:149. https://doi.org/10.1007/s11625-014-0264-6

Khairul M-M, Siahaan AP (2016) Decision support system in selecting the appropriate laptop using simple additive weighting. Int J Recent Trends Eng Res 2(12):215–222

Khan IS, Ahmad MO, Majava J (2021) Industry 4.0 and sustainable development: a systematic mapping of triple bottom line, circular economy and sustainable business models perspectives. J Clean Prod 297:126655

Krstic A, Mimovic P (2018) Measuring efficiency of the Serbian national innovation system—DEA approach. [Fifth International Scientific Conference] Contemporary Issues in Economics, Business and Management [EBM 2018], Kragujevac, pp 521–527

Krstic A, Mimovic P (2020) Evaluation and ranking of countries according to the efficiency of publishing scientific. In: XLVII simpozijum o operacionimistraživanjima, Zbornik radova. University, Faculty of Transport and Traffic Engineering, Beograd, pp 515–519

Krstić A, Mimović P, Rejman-Petrović D (2021) Integrating expert judgments into the process of measuring the efficiency of the national economy: DEA-AHP approach. Int J Quality Res 15(3):923–940. ISSN 1800-6450

Krstić A, Jakšić Mimović PM, Tadić D (2022a) Dynamic analysis of macroeconomic performance of the Balkan countries. Acta Oecon 72(3):367–391. https://doi.org/10.1556/032.2022.00027

Krstić A, Rejman-Petrović D, Nedeljković I, Mimović P (2022b) Efficiency of the use of information and communication technologies as a determinant of the digital business transformation process. Benchmark: Int J ahead-of-print No:ahead-of-print. https://doi.org/10.1108/BIJ-07-2022-0439

Kurniati E, Rahadi B, Danial M (2010) Implementing environmental management accounting (EMA) in improving eco-efficiency on corporate activities: case study on Sugarcane Company. J Appl Sci Environ Sanitation 5(4):403–410

Kurniawan R, Managi S (2017) Sustainable development and performance measurement: global productivity decomposition. Sustain Dev 25(6):639–654. https://doi.org/10.1002/sd.1684

Lira WS, Cândido GA (2008) Análise dos modelos de indicadores no contexto do desenvolvimento sustentável. Revista Perspectivas Contemporâneas 3(1):31–45

Liu F, Zhou S, Yang Y, Liu C (2022) Research on industrial ecological efficiency evaluation and improvement countermeasures based on data-driven evaluations from 30 provinces and cities in China. Sustainability 14(14):8665. https://doi.org/10.3390/su14148665

Long L (2021) Eco-efficiency and effectiveness evaluation toward sustainable urban development in China: a super-efficiency SBM–DEA with undesirable outputs. Environ Dev Sustain 23:14982–14997. https://doi.org/10.1007/s10668-021-01282-7

Lupi F, Mabkhot M, Finžgar M, Minetola P, Stadnicka D, Maffei A, Litwin P, Boffa E, Ferreira P, Podržaj P, Chelli R, Lohse N, Lanzetta M (2022) Toward a sustainable educational engineer archetype through industry 4.0. Comput Ind 134:103543, 1–16. https://doi.org/10.1016/j.compind.2021.103543

Luqmani A, Leach M, Jesson D (2017) Factors behind sustainable business innovation: the case of a global Carpet Manufacturing Company. Environ Innov Soc Trans 24:94–105

Mello JCS, Gomes EG (2004) Eficiênciasaeroportuárias: umaabordagemcomparativa com análise de envoltória de dados. Revista de Economia e Administração 3(1):15–23

Mikayilov JI, Hasanov FJ, Galeotti M (2018) Decoupling of CO_2 emissions and GDP: a time-varying cointegration approach. Ecol Ind 95:615–628

Mimović P, Krstić A, Jakšić M (2019) Dynamic analysis of the efficiency of research and development systems of South-European countries. Yugoslav J Oper Res 29(3):415–431. https://doi.org/10.2298/YJOR180815013M

Mukherjee S, Chakraborty D (2013) Is environmental sustainability influenced by socioeconomic and sociopolitical factors? Cross-country empirical evidence. Sustain Dev 21(6):353–371. https://doi.org/10.1002/sd.502

Mylonas G, Hofstaetter J, Giannakos M, Friedl A, Koulouris P (2023) Playful interventions for sustainability awareness in educational environments: a longitudinal, large-scale study in three countries. Int J Child-Comput Interact e 35:100562. https://doi.org/10.1016/j.ijcci.2022.100562

Nidumolu R, Prahalad CK, Rangaswami MR (2009) Why sustainability is now the key driver of innovation. Harv Bus Rev 87(9):56–64

Raza SA, Shah N (2018) Testing environmental Kuznets curve hypothesis in G7 countries: the role of renewable energy consumption and trade. Environ Sci Pollut Res Int 25(27):26965–26977. PMID:30008164

Rejman-Petrovic D, Krstic A, Nedeljković I, Mimovic P (2022) Efficiency of digital business transformation in the Republic of Serbia. VINE J Inf Knowl Manag Syst ahead-of-print No:ahead-of-print. https://doi.org/10.1108/VJIKMS-12-2021-0292

Rodríguez MC, Hascic I, Souchier M (2018) Environmentally adjusted multifactor productivity: methodology and empirical results for OECD and G20 countries. Ecol Econ 153:147–160

Santana NJ, Rebellato DAN, Périco AE, Mariano EB (2014) Sustainable development in the BRICS countries: an efficiency analysis by data envelopment. Int J Sust Dev World 21(3):259–272. https://doi.org/10.1080/13504509.2014.900831

Schiliro D (2019). Sustainability, innovation, and efficiency: a key relationship. In: Financing sustainable development, Chapter: IV. Palgrave, Project: Global Green Economy, pp 1–15

Sharma R, Jabbour CJC, de Sousa Jabbour ABL (2020) Sustainable manufacturing and industry 4.0: what we know and what we don't. J Enterp Inf Manag 34(1):230–266. https://doi.org/10.1108/JEIM-01-2020-0024

Soares de Mello JC, Gomes EG, Meza LA, Leta FR (2008) DEA advanced models for geometric evaluation of used Lathes. WSEAS Trans Syst 7(5):500–520

Song W, Cao S, Du M, Lu L (2023) Distinctive roles of land-use efficiency in sustainable development goals: an investigation of trade-offs and synergies in China. J Clean Prod 382:134889. https://doi.org/10.1016/j.jclepro.2022.134889

Tian N, Tang S, Che A, Wu P (2019) Measuring regional transport sustainability using super-efficiency SBM-DEA with weighting preference. J Clean Prod 242:118474. https://doi.org/10.1016/j.jclepro.2019.118474

Toma P, Miglietta PP, Zurlini G, Valente D, Petrosillo I (2017) A non-parametric bootstrap-data envelopment analysis approach for environmental policy planning and management of agricultural efficiency in EU countries. Ecol Ind 83:132–143. https://doi.org/10.1016/j.ecolind.2017.07.049

Tone K (2001) A slacks: based measure of efficiency in data envelopment analysis. Eur J Oper Res 130(3):498–509. https://doi.org/10.1016/S0377-2217(99)00407-5

Tsaples G, Papathanasiou J (2021) Data envelopment analysis and the concept of sustainability: a review and analysis of the literature. Renew Sustain Energy Rev 138. https://doi.org/10.1016/j.rser.2020.110664

Türkoglu SP, Kardogan PSO (2017) The role and importance of energy efficiency for sustainable development of the countries. International sustainable buildings Symposium. Springer, Cham, pp 53–60

Unruh G, Ettenson R (2010) Growing Green: three smart paths to developing sustainable products. Harvard Bus Rev 94–100

Verma P, Raghubanshi AS (2018) Urban sustainability indicators: challenges and opportunities. Ecol Ind 93:282–291

Viglia S, Civitillo DF, Cacciapuoti G, Ulgiati S (2017) Indicators of environmental loading and sustainability of urban systems: an emergy-based environmental footprint. Ecol Ind 94:82–99. https://doi.org/10.1016/j.ecolind.2017.03.060

Wang K, Wei Y, Huang Z (2018) Environmental efficiency and abatement efficiency measurements of China's thermal power industry: a data envelopment analysis based materials balance approach. Eur J Oper Res 269(1):35–50. https://doi.org/10.1016/j.ejor.2017.04.053

Wang X, Ding H, Liu L (2019) Eco-efficiency measurement of industrial sectors in China: a hybrid super-efficiency DEA analysis. J Clean Prod 229:53–64

Wang Z, Fu H, Liu H, Liao C (2023) Urban development sustainability, industrial structure adjustment, and land use efficiency in China. Sustain Cities Soc 89:104338. https://doi.org/10.1016/j.scs.2022.104338

Wei C, Meng J, Zhu L, Han Z (2023) Assessing progress towards sustainable development goals for Chinese urban land use: a new cloud model approach. J Environ Manag 326(Part B):116826. https://doi.org/10.1016/j.jenvman.2022.116826

Yılmaz B, Harmancıoğlu NB (2010) Multi-criteria decision making for water resource management: a case study of the Gediz River Basin, Turkey. Water SA 36(5):563–576

Zhang B, Bi J, Fan Z, Yuan Z, Ge J (2008) Eco-efficiency analysis of industrial system in China: a data envelopment analysis approach. Ecol Econ 68(1–2):306–316. https://doi.org/10.1016/j.ecolecon.2008.03.009

Zhang Z, Zhu H, Zhou Z (2022) Kai Zou How does innovation matter for sustainable performance? Evidence from small and medium-sized enterprises. J Bus Res 153:251–265. https://doi.org/10.1016/j.jbusres.2022.08.034

Zielinska-Chmielewska A, Olszanska A, Kaźmierczyk J, Vladimirovna Andrianova E (2021) Advantages and constraints of eco-efficiency measures: the case of the polish food industry. Agronomy 11:299. https://doi.org/10.3390/agronomy11020299

Zakari A, Khan I, Tan D, Aldorado R, Dagar,V (2022) Energy efficiency and sustainable development goals (SDGs). Energy 239:122365. https://doi.org/10.1016/j.energy.2021.122365

Zhou H, Yang Y, Chen Y, Zhu J (2018) Data envelopment analysis application in sustainability: the origins, development and future directions. Eur J Oper Res: EJOR (Elsevier, Amsterdam) 264:1–16. https://doi.org/10.1016/j.ejor.2017.06.023

Zhou Y, Sirisrisakulchai J, Liu J, Sriboonchitta S (2019) Factors affecting carbon emissons in the G7 and BRICS countries: evidence from quantile regression. In: 7th international symposium, IUKM, pp 406–417

Zhou XD, Xue ZK, Seydehmet J (2021) An Empirical study on industrial eco-efficiency in arid resource exploitation region of northwest China. Environ Sci Pollut Res 28:53394–53411

Zurano-Cervelló P, Pozo C, Mateo-Sanz JM, Jiménez L, Guillén-Gosálbez G (2019) Sustainability efficiency assessment of the electricity mix of the 28 EU member countries combining data envelopment analysis and optimized projections. Energy Policy 134:110921. https://doi.org/10.1016/j.enpol.2019.110921

Internet Sources

https://sustainabledevelopment.un.org/content/documents/5987our-common-future.pdf

https://data.worldbank.org/

https://ec.europa.eu/eurostat

Data-Policy Interaction

The research starts from the three dimensions on which sustainable development rests, economic, ecological, and social, which represent key points of reference in the sustainable development agenda of the United Nations from 2015 and have been accepted by the European Union. In accordance with the European Union's response to the United Nations 2030 agenda, which envisages a strategic approach to the implementation of sustainable development goals both at the level of the member states and at the level of all other interested parties, the focus of the analysis was the countries that gravitate to the Western Balkans and Eastern Europe, the latter of which have since become members of the European Union. The idea is to place the space for the implementation of sustainable development goals in the context of the need for further enlargement of the EU, which helps its sustainability in the long term, from which, implicitly, its members will also benefit. All the more so, as the countries of the Western Balkans, both geographically and politically and economically, form an inseparable part of the European space. The analysis unequivocally established the key link between membership in the European Union and the performance of sustainable development in terms of the efficiency of the realization of the goals of sustainable development and indicated the need to continue the process of association with the aim of general sustainability.

Predrag Mimović is Full Professor at the Faculty of Economics, University of Kragujevac, Serbia. His areas of research interest are operations research, decision-making, and multi-criteria decision-making. He has published several scientific papers in journals referenced in the WOS and SCOPUS databases. Also, he was Reviewer in several prominent scientific journals such as European Journal of Operations Research, Economic Research, Land Use Policies, Management Decision, Waste Management, International Journal of Information Technology and Decision Making,

Benchmarking, International Journal of Performance and Productivity Management, and Journal of Economic Studies.

Ana Krstić Srejović is Ph.D. student at the Faculty of Economics, University of Belgrade, and Teaching Assistant at the Faculty of Economics, University of Kragujevac, Serbia. Her research focus is operational research, financial and actuarial mathematics, and quality management with multi-criteria analysis, mathematical programming, and data analytics.

Milena Jakšić is Full Professor at the Faculty of Economics, University of Kragujevac, Serbia. Here scientific interests are financial system, the financial markets, the financial institutions, the measurement of the financial institutions performances, and financial institutions risk management. She is Author and Co-author of numerous academic articles and participated in national and international scientific projects and conferences.

Chapter 7
Smart Participation for Smarter Cities: Smart Citizens' Capacity Framework

Ivana Simić

Abstract Based on a holistic approach that acknowledges the diverse components critical for the efficient functioning of smart cities, this paper highlights the central role of citizens in fostering smart participation. It underscores the imperative for systematic and meaningful engagement of citizens across various aspects of city operations. In addition to a favorable environment, citizens' active involvement significantly depends on their personal capacities and competencies. However, the existing literature has insufficiently addressed the necessary aspects for individuals to become well-informed and engaged participants in urban settings. This theoretical study aims to fill this gap by introducing the Smart Citizens' Capacity Framework, an overarching structure encompassing specific competencies (a total of 14) and competency groups (4) essential for fostering smarter citizens' participation. The list of competencies presented within the Smart Citizens' Capacity Framework is not exhaustive or unchangeable and is primarily focused on those competencies that individuals can autonomously develop and refine. By emphasizing the significance of the capacities and competencies of smart citizens, this paper aims to encourage the enhancement of these abilities. This enhancement can benefit not only smart cities but also other urban centers and society at large.

Keywords Holistic approach · Smart participation · Personal capacities · Citizens' competencies · Smart cities

7.1 Introduction

Characterized as "a specific form of social space" (Barceló et al. 2002, p. 63) where large populations reside and engage in economic activities (United Nations 2016, p. 1), cities boast a history spanning centuries (Rowe 1900, p. 721). The recent surge in urbanization has led scholars to designate the twenty-first century as the

I. Simić (✉)
University of Niš, Niš, Serbia
e-mail: ivana.simic@eknfak.ni.ac.rs

J. J. Stanković et al. (eds.), *Regional and Urban Challenges and Opportunities for Data-Policy Interactions*, The Urban Book Series,
https://doi.org/10.1007/978-3-032-08592-4_7

era of "urban concentration" (McGranahan and Satterhwaite 2014, p. 4), signifying the absolute dominance of cities. Despite occupying a mere 3% of the Earth's land surface (European Commission 2020, p. 5), urban areas are expected to accommodate a substantial portion of the global population by 2050, with projections suggesting that two-thirds of the world's inhabitants, equivalent to 7 out of 10 individuals, will call cities their home (World Bank Group 2021, p. 9).

The dense concentration of people within a limited geographic space has created considerable advantages. Nevertheless, it has also engendered many challenges contemporary cities and their inhabitants face. Primary issues encompass pollution, congestion, inadequate infrastructure, waste management, high energy demands, unemployment, crime, insecurity, and an aging population, among others (Patrão et al. 2020, p. 1117; Sharifi 2019, p. 1272).

Effectively addressing the aforementioned challenges and many other urban issues necessitates harnessing knowledge from diverse scientific disciplines (Xu and Geng 2019, p. 137). Particularly, Computer Science stands out as the primary driving force (Patrão et al. 2020, p. 1118), due to its pivotal role in fostering numerous technological advancements in recent decades (UNECE 2020, p. 13).

The increasing adoption and utilization of digital technology, aimed at mitigating urban challenges and optimizing city operations, has accelerated the emergence of the concept of smart cities (OECD 2020a, p. 8). This notion swiftly evolved into a metaphor for urban modernity (Glasmeier and Christopherson 2015, p. 3), symbolizing the fusion of advanced technologies and innovative approaches within the urban environment.

7.2 A Holistic Approach to Smart Cities

The research on smart cities is primarily associated with the 1990s (Zhao et al. 2021, p. 1; Yigitcanlar et al. 2018, p. 146). Carlos Patrão and his colleagues offer a specific reference, tracing the beginning of the Smart City concept to 1992 with the publication of "The Technopolis Phenomenon: Smart Cities, Fast Systems, Global Networks", edited by David V. Gibson, George Kozmetsky, and Raymond W. Smilor (Patrão et al. 2020, p. 1118). The interest of theorists in the concept of smart cities heightened around 2010 (Ojo et al. 2016, p. 24). Renata Dameri and Annalisa Cocchia propose that this increased interest originated when the European Union began using the term "smart" to describe sustainability projects and actions in urban spaces (Dameri and Cocchia 2013, p. 4).

The attention given by theorists and researchers to the concept of smart cities has not resulted in a singular, universally accepted definition of this phenomenon (Zhao et al. 2021, p. 2; Mora et al. 2017, p. 3; Patrão et al. 2020, p. 1118; Sharifi 2019, p. 1270). While the reasons for this lack of consensus may differ, an examination of various definitions of smart cities leads to the conclusion that among the key factors are:

- Efforts by certain researchers (Giffinger et al. 2007, p. 11; Bosch et al. 2017, p. 7; Mora and Deakin 2019, p. 19) and institutions (*European Commission**; OECD 2020b, p. 8) to accentuate specific categories (characteristics, components, dimensions, resources, goals, etc.) they consider crucial for precise determination of a smart city.
- The use of specific terms (e.g., "digital city," "cyberville," "electronic city," "flexicity," "information city," "telicity," "wired city," "intelligent city," "eco-city," "green city," "knowledge city," "sustainable city," "livable cities," "resilient cities," "low carbon cities," and even combinations such as "low carbon eco cities" and "ubiquitous ecocities") (Saraju et al. 2016, p. 60; Bull and Azennoud 2016, p. 4; De Jong et al. 2015, p. 26), as synonyms for smart cities, that do not fully capture the essence of the smart city concept (Moura and De Abreu e Silva 2019, p. 1).
- The ongoing evolution of the smart city concept (UNECE 2020, p. 4), resulting in a shift in focus of interest among theorists and researchers in this field (Meijer and Bolívar 2016, p. 392).

Without diminishing the significance of any potential causes of disunity among theorists, the subsequent text focuses on the last mentioned factor. Insight into this factor facilitates understanding prevailing urban tendencies and how they are interpreted in theory.

The progression of the smart city concept is particularly evident across its developmental phases. The United Nations Economic Commission for Europe (UNECE) outlines three phases characterizing the evolution of this concept (UNECE 2020, p. 4):

- Smart City 1.0 phase—characterized by a top-down framework, with a primary focus on ICT infrastructure and the implementation of solutions promoted by technological companies.
- Smart City 2.0 phase—emphasizing a people-focused and user-friendly framework, prioritizing the needs and experiences of citizens.
- Smart City 3.0 phase—embracing an inclusive and participatory urban governance framework, while still leveraging technology to enhance city operations.

The particular elements (technology, people, and governance) highlighted within these phases, align with the dimensions of smart cities emphasized by certain theorists. For instance, Taewoo Nam and Theresa A. Pardo delineate pivotal dimensions of smart cities as follows (Nam and Pardo 2011, pp. 285–286):

- Technology dimension (determined by "technology factors": physical infrastructure, smart technologies, mobile technologies, virtual technologies, digital networks)
- Human dimension (determined by "human factors": human infrastructure, social capital)
- Institutional dimension (determined by "institutional factors": governance, policy, regulations/directives).

The evolution of the smart cities concept mentioned above suggests a transition in focus among theorists and researchers, shifting from one dimension or category of smart cities to another. By their analysis of various definitions of the smart city, Albert Meijer & Manuel Pedro Rodríguez Bolivar have identified four prevailing focal areas among theorists (see: Meijer and Bolívar 2016, pp. 396–397):

- Technological focus (technology is the central characteristic or dimension of the smart city).
- Human resource focus (people are the key feature of smart cities).
- Governance focus (collaboration and interaction among various stakeholders in the city are the defining features of the smart city).
- Combined focus (the combination of smart technology, smart people, and smart collaboration in the city is essential for the smart city).

The combined focus stresses a holistic approach to smart cities (Rozario et al. 2021, p. 1), advocated by theorists emphasizing the necessity for a comprehensive interpretation and understanding of smart urban environments (Breuer et al. 2014; Zhao et al. 2021, p. 9; Hemment and Townsend 2013, p. 2; Yigitcanlar et al. 2018, p. 157). An exemplary definition illustrating this combined focus, or a holistic perspective, is presented by Andrea Caragliu and his colleagues, emphasizing that "A city can be defined as 'smart' when investments in human and social capital, alongside traditional (transport) and modern (ICT) communication infrastructure, fuel sustainable economic development and a high quality of life, with a wise management of natural resources through participatory governance" (Caragliu et al. 2011).

The importance of embracing a holistic approach lies in the shift from the formerly dominant technological viewpoint (see: Washburn et al. 2010, p. 2) to one centered on people. This perspective does not portray individuals solely as passive recipients of urban services; instead, it recognizes them as active contributors intertwined with other relevant stakeholders. They actively participate in designing, implementing, monitoring, and evaluating of smart city initiatives, exerting considerable influence over its operations and outcomes. As Catherine Mulligan emphasizes: Without engaging citizens about the role of technology in their cities and its impact on the way in which they will travel, live, eat, play, and study, the smart city vision will fail… A smart city, therefore, starts with smart citizens who are asked their opinions and engaged in the process of deciding how they are used" (Mulligan 2013, p. 84).

7.3 Smart Citizens' Participation

The recognition that the potential of smart cities extends beyond technology alone and that achieving smart city status requires systematic and meaningful engagement of its citizens has been identified by the UNECE (2020, p. xiii). In this context, "systematic and meaningful engagement" refers to the active participation of citizens in fulfilling their right to the city. This entails a focus on promoting prosperity, equity, comfort, and innovation in cities, addressing social needs, and ensuring that housing and urban services are of high quality and affordable. It also involves catering to the needs of vulnerable populations and individuals with disabilities, adopting a gender-sensitive and age-responsive approach, and recognizing residents' diverse and evolving needs at different stages of life (UNECE 2020, p. 3).

Many contemporary theorists have also defined participation, emphasized its importance and explained its role in the functioning of smart cities. For example, Benoit Granier and Hiroko Kudo define participation as the involvement of citizens in the design, implementation, monitoring, and evaluation of public policies (Granier and Kudo 2016, p. 61, 63). Antony Simonofski and his colleagues point out that although "smart cities can provide innovative solutions in various domains such as environment, economy, mobility, safety with technology as enabler… this is only possible if the citizens, the end-users, are involved in the design of smart city" (Simonofski et al. 2019, p. 47). Taylor Shelton and Thomas Lodato believe that "participation of citizens is crucial to any truly democratic mode of urban governance" (Shelton and Lodato 2019, p. 35). Citizen participation is also linked to democracy, as Michael Cuthill and John Fien note that the concept of citizen participation "is part of the wider discussion about democracy that extends over nearly 3000 years" (Cuthill and Fien 2005, p. 64).

Several recent OECD documents (OECD 2020a, 2022) have also underscored the significance of citizen participation. According to one of these documents, participatory democracy has gained momentum since the activist movements of the 1960s, driven by the growing complexity of policy-making and the need to address pressing policy issues, resulting in collective engagement among politicians, policymakers, civil society organizations, and citizens in making public decisions in the twenty-first century (OECD 2020a, p. 3–4). The document highlights that citizen participation encompasses various forms of involvement in the policy cycle, as well as in service design and delivery. As the OECD (2022) states, participation refers to the efforts made by public institutions to listen to citizens' perspectives, inputs, and views, enabling them to exert influence on the activities and decisions of public authorities at different stages of the policy cycle (OECD 2022, p. 13).

In compliance with the above mentioned, this paper considers smart citizens' participation as a systematic and meaningful engagement of citizens in the design, implementation, monitoring, and evaluation of smart city operations and outcomes (compare with: UNECE 2020, xiii; Granier and Kudo 2016, p. 61).

In addressing the problems that cities face, citizens can be involved to varying degrees. Despite being introduced as early as 1969, Arnstein's ladder of citizen participation (Fig. 7.1) (Arnstein 1969), continues to be regarded as a valid framework for understanding citizen involvement in problem-solving (Schelings and Elsen 2019, p. 10).

Certain theorists (Schelings and Elsen 2019, p. 10; Cardullo and Kitchin 2019a) have expanded and revised Arnstein's typology to better align with the evolving contexts of citizen participation. Within this context, the adaptation proposed by Clémentine Schelings and Catherine Elsen, which applies Arnstein's typology to the realm of smart cities, is notably interesting (Fig. 7.2).

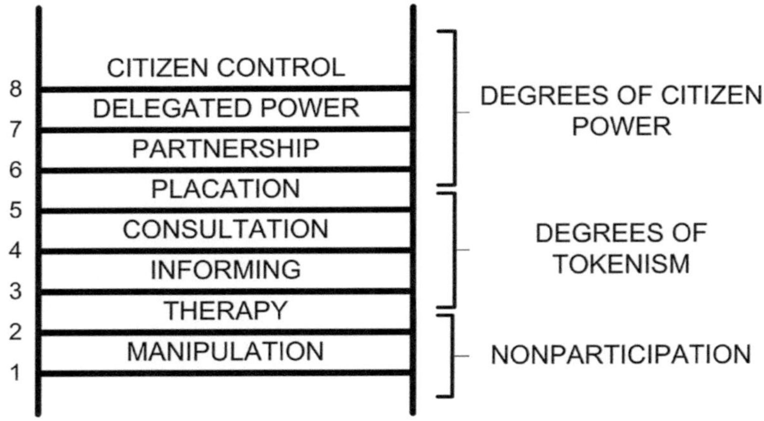

Fig. 7.1 Arnstein's ladder of citizen participation. *Source* Arnstein (1969, p. 217)

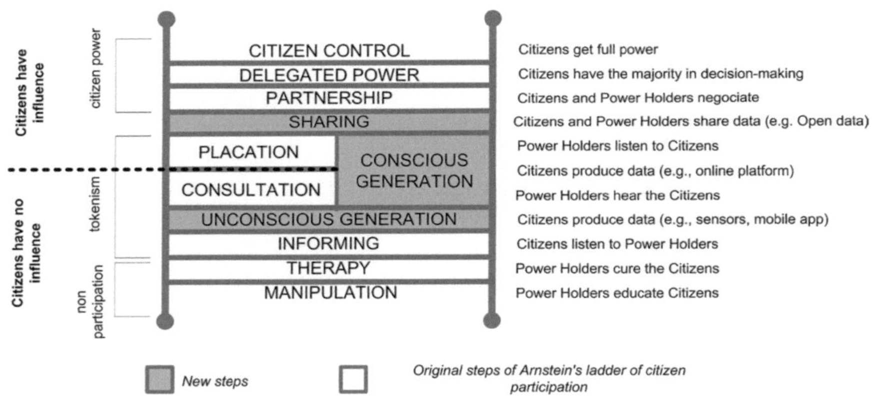

Fig. 7.2 Arnstein's ladder of citizen participation adapted to the smart city context. *Source* Schelings and Elsen (2019, p. 11)

According to Schelings and Elsen, the apex of citizen participation, termed "citizen control," entails citizens possessing complete power and engaging continuously at every stage of the process (Schelings and Elsen 2019, p. 10). Nonetheless, they acknowledge that achieving this ideal scenario might not be realistic due to time constraints (Schelings and Elsen 2019, p. 10). Paolo Cardullo and Rob Kitchin also observe that while the highest levels of citizens participation may mirror citizens' aspirations, they can be challenging to realize practically (Cardulllo and Kitchin 2019b, p. 9). In their assessment of smart city participation in Dublin, Cardullo and Kitchin underscored that "citizens occupy a largely passive role, with companies and city administrations performing forms of civic paternalism (deciding what's best for citizens) and stewardship (delivering on behalf of citizens)" (Cardullo and Kitchin 2019a, p. 814).

From my perspective, pursuing the "ideal scenario" holds value, and under the "right" conditions, it has the potential to become the norm. The advancements in digital technology, particularly in artificial intelligence, are contributing to the creation of these conditions sooner than anticipated. However, achieving the highest level of citizen participation necessitates not only a more proactive role from smart city governments but also a heightened interest from citizens in participating and enhancing their own capacities and competencies to facilitate systematic and meaningful engagement. As highlighted in one of the OECD documents above (OECD 2022, p. 15), it is not sufficient for governments to decide to engage more with citizens or to create an environment for citizen engagement; citizens themselves must also change their behavior and mindset to actively participate (OECD 2022, p. 62, 63).

While influencing citizens' behavior to enhance their participation levels encompasses various factors (Simonofski et al. 2017; Schelings and Elsen 2019; Arnstein 1969), the subsequent text predominantly concentrates on personal capacities and competencies that citizens can personally cultivate and improve. The central research question is: What capacities and competencies should smart citizens possess, which they can independently develop, to participate actively and efficiently in smart cities?

7.4 Smart Citizens' Capacities (SCC)

7.4.1 Capacities and Competencies

Theorists focusing on capacity highlight that it is a wide-ranging, challenging, and elusive concept that "is not simple to address" (Isaza et al. 2015, p. 2; Brown et al. 2001, p. 4; Baser et al. 2008, p. 8). According to the United Nations Development Group (UNDG), capacity is defined as "the ability of people, organizations and society as a whole to manage their affairs successfully" (UNDG 2017, p. 5). Similarly, the United Nations Development Programme (UNDP) characterizes capacity as "the ability of individuals, institutions and societies to perform functions, solve problems, and set and achieve objectives in a sustainable manner" (UNDP 2010, p. 2). These definitions explicitly imply the feasibility of examining capacity across various levels: individual (micro), organizational (meso), and societal (macro) (UNDG 2017, p. 5; Baser et al. 2008, p. 13).

From an individual standpoint, capacity is defined by Cohen (1993) who associates capacity with competencies, defining it as the "individual ability, competence to carry out a specific task" (Cohen 1993, p. 2). Richard Boyle and Joanna O'Riordan also emphasize the close relationship between capacity and competencies, stating, "To a large extent, capacity is achieved through competencies…" (Boyle and O'Riordan 2013, p. 3). An interesting perspective of the relationship between capacity and competencies is articulated by Heather Baser and her colleagues, who conceptualize capacity as "an umbrella concept" that encompasses the "combination of individual competencies…" (Baser et al. 2008, p. 3).

Like capacity, competencies also lack a standardized definition in the literature (Schneider 2019; Salerno et al. 2014, p. 358). Robert W. White, who introduced the concept (Schneider 2019, p. 1938), equates competencies with capacities, defining competencies as "… an organism's capacity to interact with its environment…" (Schneider 2019, p. 1939). Guy Le Boterf describes competencies as "a recognized and proven set of representations, knowledge, skills, and attitudes pertinently mobilized and combined in a given context" (Salerno et al. 2014, p. 358). The Council of Europe's Reference Framework of Competences for Democratic Culture delineates competencies as "the ability to mobilize and deploy relevant values, attitudes, skills, knowledge, and/or understanding in order to respond appropriately and effectively to the demands, challenges, and opportunities that are presented by a given type of context" (Council of Europe 2018, p. 32). The variations in defining competencies can be ascribed, partly, to the lack of consensus among theorists concerning the fundamental dimensions. For example, the widely referenced KSA model of competency dimensions emphasizes knowledge (K), skills (S), and attitudes (A) (Salerno et al. 2014, p. 359). It is evident that numerous definitions of competencies, including those mentioned previously, encompass additional dimensions such as representations, values, and critical understanding, which go beyond the parameters listed in the KSA model.

7.4.2 Literature Review

In the literature, people's individual capacities appear as "a residual category" (Isaza et al. 2015, p. 9). A comparable trend is noticeable in literature that discusses citizens' capacities within a broader context. A few theorists tend to adopt a partial approach, focusing exclusively on specific aspects when analyzing this phenomenon. For instance, Christine Wamsler and Ebba Brink explore the concept of "adaptive capacity of citizens" concerning their response to climate change (Wamsler and Brink 2014, p. 69). Similarly, Shawn W. Rosenberg delves into the social dimensions of citizens' individual capacities (Rosenberg 2007, p. 335). By equating "capacity" and "competence," John Gaventa scrutinizes these two categories in the context of strengthening democracy and while examining various dimensions of power (Gaventa 1995, p. 29, 34).

The situation mirrors a similar trend in the literature regarding citizenship competence. Notably, Geert ten Dam and Monique Volman (Ten Dam and Volman 2004) exemplify a partial approach by focusing on "critical thinking" as an essential competency for citizens. Similarly, Vuorikari et al. (2022) underscore the significance of digital competencies among citizens. While acknowledging the undeniable value of these "partial" analyses in shaping our understanding of citizens' competencies, it is crucial to recognize the scarcity and importance of more systematic and comprehensive analyses. In this context, it is noteworthy to mention the Council of Europe's Reference Framework of Competences for Democratic Culture (RFCDC) (Fig. 7.3) (Council of Europe 2018, p. 38). This conceptual model delineates 20 competencies categorized into values, attitudes, skills, and knowledge with critical understanding. It includes the appreciation of human dignity, human rights, cultural diversity, democracy, justice, equality, and the rule of law as values. Attitudes highlighted encompass openness to cultural diversity, respect, civic-mindedness, responsibility, self-efficacy, and tolerance of ambiguity. The identified skills span autonomous learning, analytical and critical thinking, listening and observation, empathy, flexibility and adaptability, communication, cooperation, and conflict resolution. Moreover, it covers a broad spectrum of knowledge areas, including politics, law, human rights, culture, religion, history, media, economies, environment, and sustainability (compare with: Barrett 2021, p. 147).

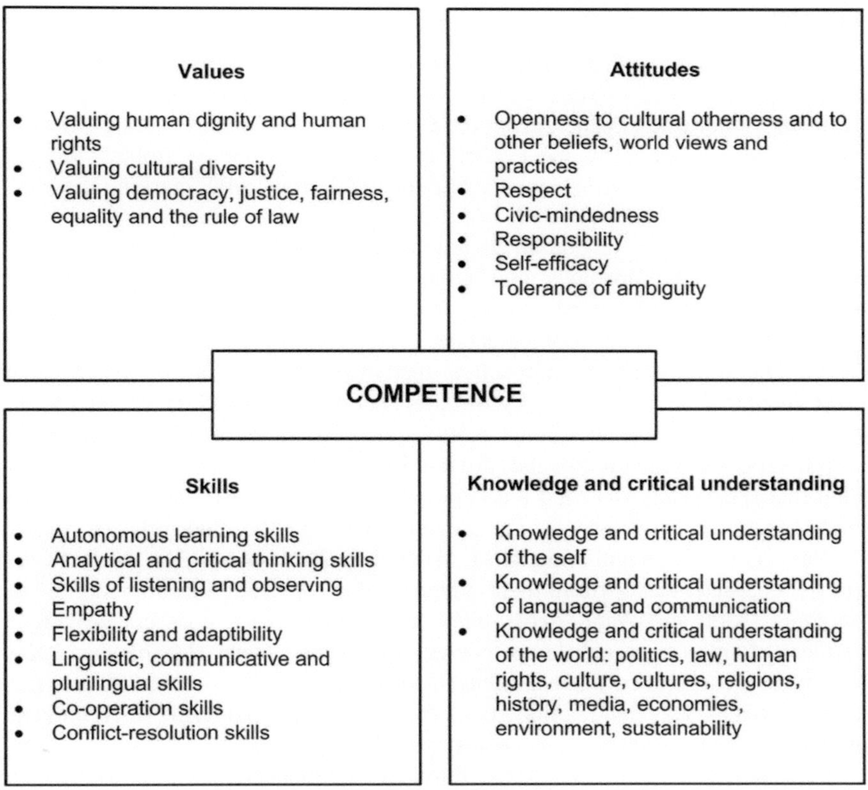

Fig. 7.3 Citizenship competencies according to the RFCDC. *Source* Council of Europe (2018, p. 38)

Gerald Steiner and his colleagues (Steiner et al. 2013) present an insightful overview of the competencies essential for citizens to effectively address the challenges encountered during their active participation. The authors highlight several key competencies, including (Steiner et al. 2013, pp. 575–576):

- Professional domain competence: specific knowledge, skills, and methods within a particular field.
- Systemic competence: understanding and dealing with complex processes, events, and systems (based on holistic system thinking).
- Sociocultural (collaborative) competence: effective communication and interaction across disciplines, organizations, and stakeholder groups, involving teams, integration, reflection, conflict management, establishing common goals, dealing with diverse cultural settings.
- Personal competence: self-awareness, self-reflection, self-motivation, autonomy, and future-oriented thinking.
- Creativity competence: generating original solutions that go beyond routine problem-solving.

To conduct an in-depth assessment of academic literature pertaining to the capacities and competencies of smart citizens, a systematic literature search was carried out by the author of this paper, using the Google Scholar database. Google Scholar is widely acknowledged as the largest academic search engine globally (López-Cózar et al. 2019). The search spanned publications from 1992, marking the initial publication on smart cities, to 2023.

To address the lack of consensus among theorists regarding terminology, the search covered both "capacities" and "competencies," along with "people" and "citizens." Subsequently, key terms such as "smart citizens' capacities," "smart people capacities," "smart citizens' competencies," and "smart people competencies" were utilized. This comprehensive approach resulted in a total of 19,900 articles for "smart citizens' capacities," 17,900 articles for "smart people capacities," 19,200 articles for "smart citizens' competencies," and 18,500 articles for "smart people competencies."

Following the utilization of the "Advanced Search" option, not one article was found that specifically matched the keywords "smart citizens' capacities," "smart people capacities," or "smart people competencies" in either their titles or content. Only two papers (Klett and Wang 2014; Mikhieieva 2021) contain the phrase "smart citizens' competencies" within their content. Upon a thorough examination of the first paper (Klett and Wang 2014), one can conclude that it does not provide an overview of smart citizens' competencies. The second paper (Mikhieieva 2021), available as a conference abstract online, does not specify the specific competencies of smart citizens. Moreover, among titles containing the exact term "smart citizens' competencies," only that conference abstract was identified.

The literature review proceeded by eliminating overlapped papers. The remaining papers were primarily analyzed based on their abstracts. In instances where the abstract indicated a potentially systematic review of individual capacities or competencies of smart citizens, the entire paper would be analyzed, provided it was accessible online.

It is interesting to note that a partial approach to the capacities and competencies of smart citizens has been identified within some papers. The prevailing trend in the literature is to analyze the capacities and competencies of smart citizens within the context of digital technology's influence, thereby emphasizing the significance of digital capacities and competencies among smart citizens (Williamson 2015, 2022; Noveck 2015; Atif et al. 2017; Schweitzer 2017, p. 4).

As valuable resources for developing a more comprehensive framework of capacities and competencies of smart citizens, a few papers which explore the concept of "civic intelligence," authored by Schuler (2014, 2016), have been identified.

According to Schuler, "civic intelligence" pertains to the ability of collectives, ranging from small informal groups to humanity as a whole, to effectively and fairly address shared problems (2014, p. 519). While Schuler suggests that the concept of civic intelligence can be applied to any community, he also emphasizes its particular relevance to smart cities and their inhabitants. Specifically, Schuler highlights that while intelligent individuals are crucial for smart cities, their intelligence should be complemented by what has traditionally been referred to as wisdom, or as Schuler puts it, civic intelligence (Schuler 2016, p. 52).

Schuler has devised a comprehensive framework (in this paper termed the "S-Framework"), encompassing essential capacities for civic intelligence. This framework (depicted in Fig. 7.4) consists of five distinct categories (knowledge, attitude and aspirations, organizational capital, relational capital, and financial and material resources) comprising a total of 26 capacities and 14 sub-capacities crucial to civic intelligence (Schuler 2014, p. 520). All these elements offer a means to comprehend better and potentially evaluate smart citizens' civic intelligence.

Upon examining the S-Framework within the context of smart cities, it becomes evident that it incorporates specific personal capacities expected of smart citizens, primarily within the "knowledge" and "attitude and aspiration" categories. However, it also includes specific capacities, such as "organizational structure," relevant to smart participation but not attributed to the individual capacities of smart citizens. Furthermore, Schuler's online papers on civic intelligence (Schuler 2001, 2009, 2010, 2012, 2014, 2016), lack a detailed and precise explanation of all the categories and capacities outlined in the S-Framework.

7.4.3 Modeling SCC Framework

Based on the aforementioned context, this paper introduces the Smart Citizens' Capacity Framework (hereinafter SCC Framework) (Table 7.1). This framework aims to outline the essential capacities and competencies necessary for active participation of smart citizens. Within the SCC Framework, capacities are conceptualized as an overarching structure encompassing various individual competencies (compare with: Baser et al. 2008, p. 3). These individual competencies are organized within the SCC Framework in alignment with the competency dimensions identified in RFCDC (Council of Europe 2018, p. 38; Barrett 2021, p. 147), the KSA model (Salerno

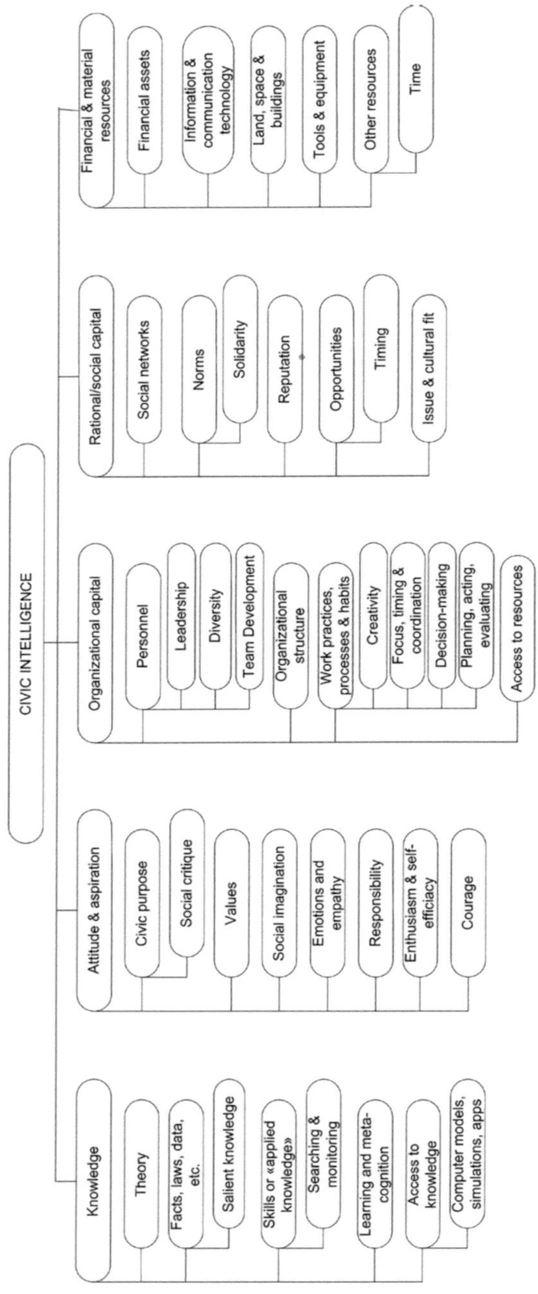

Fig. 7.4 Capacities for civic intelligence framework. *Source* Schuler (2014, p. 520)

et al. 2014, p. 359), and the S-Framework (Schuler 2014, p. 520). Consequently, specific groups or sets of individual competencies have been created within the SCC Framework, namely: knowledge, skills, attitudes and aspirations, and values.

The specific competencies within the SCC Framework were primarily established with the support of the content presented within the S-Framework. To accomplish this, the S-Framework underwent modifications and supplements. For instance, the term "competencies" (SCC Framework) is used instead of "capacities" and "sub-capacities" (S-Framework). Moreover, certain non-personal "capacities" present within S-Framework (e.g., "organizational structure") were omitted from the SCC Framework, while others were renamed (e.g., "theory" to "conceptual knowledge"; "Facts, laws, data, etc." contributed to the creation of "contextual knowledge") (Sáenz 2009). Additionally, the SCC Framework has been enriched with certain pertinent competencies relevant to smart citizens' participation, such as digital skills (Vuorikari et al. 2022; Williamson 2022; Noveck 2015; Atif et al. 2017), critical thinking skills (Ten Dam and Volman 2004; Barrett 2021, p. 147), and problem-solving skills (Steiner et al. 2013; Yang 2012). "Values" have been distinguished as a separate group of competencies within the SCC Framework and classified based on the categorization of positive values proposed by Jan Cieciuch and his colleagues (Cieciuch et al. 2015, p. 43). Furthermore, a concise explanation on each distinct individual competence within the SCC Framework is provided, supported by relevant literature references.

7.4.4 Limitations and Future Research

The list of capacities and competencies delineated within the SCC Framework primarily relies on theoretical rather than empirical research, representing a limitation of this study. Additionally, despite using the Google Scholar database, the largest academic search engine available, a limited number of comprehensive papers specifically addressing the capacities and competencies of smart citizens were identified. This restricted access to resources crucial for the development of the SCC Framework. It is also important to note that while the content of the SCC Framework seems logical and well-structured, it should not be considered definitive or unchangeable.

Considering the rapid evolution of cities and society in general, future research focusing on both conceptual and empirical studies concerning the capacities and competencies essential for enabling active engagement of smart citizens will be crucial. Validating, expanding, and refining the contents of the SCC Framework through such research endeavors is imperative to ensure its continued relevance and applicability.

Table 7.1 Smart citizens' capacity framework (SCC Framework)

Smart citizens' capacities for smarter participation	Competency groups	Competencies
	Knowledge	**Conceptual knowledge**—understanding and familiarity with the relevant concepts, theories, and principles related to the problem or issue in which smart citizens are participating. It is typically defined as "…knowledge that is rich in relationships. It can be thought of as a connected web of knowledge, a network in which the linking relationships are as prominent as the discrete pieces of information. Relationships pervade the individual facts and propositions so that all pieces of information are linked to some network" (Star and Stylianides 2013) **Contextual knowledge**—understanding the environment (social, political, cultural, economic) in which the problem is situated. It is directly "related to everyday-life problems in the real world" (Sáenz 2009, p. 126) **Salient knowledge**—understanding concealed information relevant to the specific issue. More precisely, it encompasses "access to local knowledge of the constituencies, opponents, and third parties with which one is interacting" (Ganz 2005, p. 220)
	Skills	**Learning skills**—the ability to learn and acquire new knowledge, skills, and competencies needed to actively participate in smart cities; being able to continuously improve and develop oneself (see more in: Higgins et al. 2007). Learning is often defined as "a relatively permanent change in behavior, knowledge and thinking skills as a result of experience" (Nagel and Scholes 2017, p. 11) **Critical thinking skills**—the ability to analyze and evaluate information objectively, logically, and systematically. It involves also questioning assumptions, considering multiple perspectives, and drawing evidence-based conclusions (see more in: Heard et al. 2020, p. 2) **Problem-solving skills**—"… a collection of small component skills as well as strategies for managing them… directed at transforming a given situation into a goal situation when no obvious solution method is available to the problem solver" (Mayer 1999, p. 445, 437); "the ability to find causes, find solutions, and avoid problems" (see: Yang 2012, p. 366); "higher order thinking skills such as 'visualization, association, abstraction, comprehension, manipulation, reasoning, analysis, synthesis, generalization'" (Foshay and Kirkley 1998, p. 3) **Digital skills**—"the ability to use, access, filter, evaluate, create, program and share digital content" (The Council of European Union 2018, p. 9)

(continued)

Table 7.1 (continued)

	Attitudes and aspirations	**Civic Purpose**—refers to an individual's feeling of responsibility and commitment to the well-being of community or broader society. It requires recognizing the common good and actively participating in beneficial activities and initiatives for community's benefit, such as volunteering, engaging in community projects, voting, etc. It is defined as "a sustained intention to contribute to the world beyond the self through civic or political action" (Malin et al. 2015, p. 109) **Social Imagination**—"to imagine a situation in which the situation has been changed for the better" (Schuler 2014, p. 523) **Emotions and Empathy**—"to feel in some way what other people are feeling" (Schuler 2014, p. 524) **Ambition**—"the persistent and generalized striving for success, attainment, and accomplishment" (Judge and Kammeyer-Mueller 2012) **Enthusiasm**—a strong and eager interest or passion for something **Courage**—the willingness of citizens to openly express their views and ideas, to freely criticize what deserves criticism, to set ambitious goals (Baltzersen 2022, p. 411)
	Values	**Positive values**—"transsituational goals varying in importance, that serve as guiding principles in the life of a person or group" (Cieciuch et al. 2015, p. 43). Some of the most prominent values are (Cieciuch et al. 2015, p. 44) • **universalism** (understanding, appreciation, tolerance, and protection for the welfare of all people and for nature) • **benevolence** (preservation and enhancement of the welfare of people with whom one is in frequent personal contact) • **conformity** (restraint of actions, inclinations, and impulses likely to upset or harm others and violate social expectations or norms) • **security** (safety, harmony and stability of society, of relationships, and of self) • **achievement** (personal success through demonstrating competence according to social standards) • **stimulation** (excitement, novelty, and challenge in life)

Source Author's own

7.5 Conclusion

Although the capacities and competencies crucial for enabling active and informed participation in society are important, they have yet to receive adequate attention in the literature. This paper aims to bridge this gap by introducing the SCC Framework, serving as an overarching structure encompassing specific competencies (total of 14) categorized into four groups: knowledge (conceptual, contextual, salient), skills (learning, critical thinking, problem-solving, digital), attitudes and aspirations (civic purpose, social imagination, emotions and empathy, ambition, enthusiasm, courage), and values (positive values).

Regarding these competencies, two key points are essential to note. Firstly, the list of competencies presented within the SCC Framework is not exhaustive, and ongoing research on smart citizens' capacities may lead to further refinement. Secondly, while these competencies are analyzed within the context of smart cities, their relevance extends beyond urban environments to various contemporary settings.

References

Arnstein SR (1969) A ladder of citizen participation. J Am Plann Assoc 35(1):216–224

Atif Y, Sergis S, Sampson D, Mathiason G (2017) A Cyberphysical learning approach for digital smart Citizenship competence development. In: WWW '17 companion: Proceedings of the 26the international conference on World Wide Web Companion, pp 397–405. https://doi.org/10.1145/3041021.3054167

Baltzersen RK (2022) Cultural-historical perspectives on collective intelligence: patterns in problem solving and innovation. Cambridge University Press, Cambridge. https://doi.org/10.1017/9781108981361

Barceló JA, Pelfer G, Mandolesi A (2002) The origins of the city: From social theory to archeological description. Archeologia e Calcolatori 13:41–63

Barrett M (2021) Citizenship competences. Scuola democratica, Maggio, pp 145–160. https://doi.org/10.12828/100675

Baser H, Morgan P, Bolger J, Brinkerhoff D, Land A, Taschereau S, Watson D, Zinke J (2008) Capacity, change and performance: study report. European Centre for Development Policy Management. https://ecdpm.org/work/capacity-change-and-performance-study-report

Bosch P, Jongeneel S, Rovers V, Neumann HM, Airaksinen M, Huovila A (2017) CITY keys indicators for smart city projects and smart cities. Technical Report. https://doi.org/10.13140/RG.2.2.17148.23686. https://eko.zagreb.hr/UserDocsImages/arhiva/Slike/projekt%20CITYKeys/rezultati/CITYkeysD14Indicatorsforsmartcityprojectsandsmartcities.pdf

Boyle R, O'Riordan J (2013) Capacity and competency requirements in local Government. Institute of Public Administration. https://www.ipa.ie/_fileUpload/Documents/CAPACITYandCOMPETENCY_SEPT2013.pdf

Breuer J, Walravens N, Ballon P (2014) Beyond defining the smart city: Meeting top-down and bottom-up approaches in the middle. TeMa J Land Use Mob Environ, Special Issue:153–164

Brown L, LaFond A, Macintyre K (2001) Measuring capacity building. Carolina Population Center, University of North Carolina at Chapel Hill. https://pdf.usaid.gov/pdf_docs/PNACM119.pdf

Bull R, Azennoud M (2016) Smart citizens for smart cities: participating in the future. Proc Inst Civil Eng Energy 169(3):93–101 (Themed issue on the smart future). https://doi.org/10.1680/jener.15.00030

Caragliu A, Del Bo C, Nijkamp P (2011) Smart cities in Europe. J Urban Technol 18(2):65–82. https://doi.org/10.1080/10630732.2011.601117

Cardullo P, Kitchin R (2019a) Smart urbanism and smart citizenship: the neoliberal logic of 'citizen focused' smart cities in Europe. EPC: Politics Space 37(5):813–830

Cardullo P, Kitchin R (2019b) Being a 'citizen' in the smart city: up and down the scaffold of smart citizen participation in Dublin, Ireland. GeoJournal 84:1–13. https://doi.org/10.1007/s10708-018-9845-8

Cieciuch J, Schwartz SH, Davidov E (2015) Values, social psychology of. In: Wright JD (ed) International encyclopedia of the social & behavioral sciences, 2nd edn., vol 25. Elsevier, Oxford, pp 41–46. https://www.researchgate.net/publication/327534758_Values

Cohen JM (1993) Building sustainable public sector managerial, professional, and technical capacity: a framework for analysis and intervention. (Development Discussion Papers, n. 473). Harvard Institute for International Development, Harvard University

Council of Europe (2018) Reference framework of competences for democratic culture. Council of Europe Publishing. https://rm.coe.int/prems-008318-gbr-2508-reference-framework-of-competences-vol-1-8573-co/16807bc66c

Cuthill M, Fien J (2005) Capacity building: facilitating citizen participation in local governance. Aust J Public Adm 64(4):63–80

Dameri RP, Cocchia A (2013) Smart city and digital city: twenty years of terminology evolution. In: Proceedings of the X conference of the Italian chapter of AIS, ITAIS, Milan, Italy, pp 1–8. http://www.itais.org/proceedings/itais2013/pdf/119.pdf

De Jong M, Jos S, Schraven D, Zhan C, Weijnen M (2015) Sustainable-smart-resilient-low carbon-eco-knowledge cities; making sense of a multitude of concepts promoting sustainable urbanization. J Clean Prod 109(16):25–38. https://doi.org/10.1016/j.jclepro.2015.02.004

*European Commission**. What are smart cities? https://commission.europa.eu/eu-regional-and-urban-development/topics/cities-and-urban-development/city-initiatives/smart-cities_en

European Commission (2020) Proposed mission: 100 climate-neutral cities by 2030—By and for Citizens. Publications Office of the European Union, Luxembourg

Foshay R, Kirkley J (1998) Principles for teaching problem solving. TRO Learning Inc., Edina

Ganz M (2005) Why David sometimes wins: strategic capacity in social movements. In: Messick DM, Kramer RM (eds) The psychology of leadership: new perspectives and research. Lawrence Erlbaum Associates, Inc., New Jersey, pp 209–238

Gaventa J (1995) Citizen knowledge, citizen competence and democracy building. Good Soc 5(3):28–35

Giffinger R, Fertner C, Kramar H, Kalasek R, Pichler-Milanović N, Mejers E (2007) Smart cities: ranking of European medium-sized cities. Vienna UT, Vienna

Glasmeier A, Christopherson S (2015) Thinking about smart cities. Camb J Reg Econ Soc 8:3–12. https://doi.org/10.1093/cjres/rsu034

Granier B, Kudo H (2016) How are citizens involved in smart cities? Analysing citizen participation in Japanese "Smart Communities." Inf Polity 21(1):61–76

Heard J, Scoular C, Duckworth D, Ramalingam D, Teo I (2020) Critical thinking: definition and structure. Australian Council for Educational Research. https://research.acer.edu.au/ar_misc/38

Hemment D, Townsend A (2013) Here come the smart citizens. In: Hemment D, Townsend A (eds) Smart citizens. FutureEverything, pp 1–4

Higgins S, Baumfield V, Hall E (2007) Learning skills and the development of learning capabilities. Technical Report. EPPI—Centre, Social Science Research Unit, Institute of Education, University of London, London

Isaza C, Herrera P, Lozano JC, Mendez K, Balanzo A (2015) Capacity: a literature review and a research agenda. https://doi.org/10.2139/ssrn.2824486

Judge TA, Kammeyer-Mueller JD (2012) On the value of aiming high: the causes and consequences of ambition. J Appl Psychol 97(4):758–775. https://doi.org/10.1037/a0028084

Klett F, Wang M (2014) Editorial: Smart cities of the future: creating tomorrow's education toward effective skills and career development today. Knowl Manag E-Learn 6(4):344–355

López-Cózar ED, Orduna-Malea E, Martín-Martín A (2019) Google scholar as a data source for research assessment. In: Glänzel W, Moed HF, Schmoch U, Thelwall M (eds) Springer handbook of science and technology indicators. Springer Handbooks. Springer, Cham, pp 95–127. https://doi.org/10.1007/978-3-030-02511-3_4

Malin H, Ballard PJ, Damon W (2015) Civic purpose: an integrated construct for understanding civic development in adolescence. Hum Dev 58(2):103–130. https://doi.org/10.1159/000381655

Mayer RE (1999) Problem solving. In: Encyclopedia of creativity, vol 2, pp 437–447

McGranahan G, Satterthwaite D (2014) Urbanisation concepts and trends. IIED Working Paper. IIED, London

Meijer A, Bolívar MPR (2016) Governing the smart city: a review of the literature on smart urban governance. Int Rev Adm Sci 82(2):392–408. https://doi.org/10.1177/0020852314564308

Mikhieieva O (2021) Smart competences for smart citizens. In: 2021 IEEE European technology and engineering management summit (E-TEMS), pp 127–130. https://ieeexplore.ieee.org/document/9524859

Mora L, Bolici R, Daekin M (2017) The first two decades of smart-city research: a bibliometric analysis. J Urban Technol 24(1):3–27

Mora L, Deakin M (2019) Untangling smart cities: from utopian dreams to innovation systems for a technology-enabled urban sustainability. Elsevier

Moura F, De Abreu e Silva JA (2019) Smart cities: definitions, evolution of the concept and examples of initiatives. In: Filho WL et al (eds) Industry, innovation and infrastructure, encyclopedia of the UN sustainable development goals. Springer Nature, Switzerland, pp 1–9. https://doi.org/10.1007/978-3-319-71059-4_6-1

Mulligan C (2013) Citizen engagement in smart cities. In: Hemment D, Townsend A (eds) Smart citizens. FutureEverything, pp 83–86

Nagel M, Scholes L (2017) Understanding development and learning: implications for teaching. Oxford University Press

Nam T, Pardo TA (2011) Conceptualizing smart city with dimensions of technology, people, and institutions. In: The proceedings of the 12th annual international conference on digital government research, Dg.o'. College Park, MD, USA, pp 282–291. https://doi.org/10.1145/2037556.2037602

Noveck BS (2015) Smart citizens, smarter state: the technologies of expertise and the future of governing. Harvard University Press

OECD (2020a) Innovative citizen participation and new democratic institutions: catching the deliberative wave. https://www.oecd.org/gov/open-government/innovative-citizen-participation-new-democratic-institutions-catching-the-deliberative-wave-highlights.pdf

OECD (2020b) Smart cities and inclusive growth. https://www.oecd.org/cfe/cities/OECD_Policy_Paper_Smart_Cities_and_Inclusive_Growth.pdf

OECD (2022) OECD guidelines for citizen participation processes. OECD Publishing, Paris. https://doi.org/10.1787/f765caf6-en

Ojo A, Dzhusupova Z, Curry E (2016) Exploring the nature of the smart cities research landscape. In: Gil-Garcia J, Pardo T, Nam T (eds) Smarter as the new urban agenda. Public Administration and Information Technology, vol 11. Springer, pp 23–47. https://doi.org/10.1007/978-3-319-17620-8_2

Patrão C, Moura P, de Almeida AT (2020) Review of smart city assessment tools. Smart Cities 3:1117–1132. https://doi.org/10.3390/smartcities3040055

Rosenberg SW (2007) Rethinking democratic deliberation: the limits and potential of citizen participation. Polity 39(3):335–360

Rowe LS (1900) The city in history. Am J Sociol 5(6):721–745

Rozario SD, Venkatraman S, Marimuthu M, Sadegh Khaksar SM, Subramani G (2021) Creating smart cities: a review for holistic approach. Appl Syst Innov 4(7):1–24. https://doi.org/10.3390/asi4040070

Sáenz C (2009) The role of contextual, conceptual and procedural knowledge in activating mathematical competencies (PISA). Educ Stud Math 71:123–143. https://doi.org/10.1007/s10649-008-9167-8

Salerno S, Nunziante A, Santoro G (2014) Competences and knowledge: key-factors in the smart city of the future. Knowl Manag E-Learn 6(4):356–376

Saraju MP, Uma C, Elias K (2016) Everything you wanted to know about smart cities. IEEE Consum Electron Mag 5(3):60–70. https://doi.org/10.1109/MCE.2016.2556879

Schelings C, Elsen C (2019) "Smart" participation: confronting theoretical and operational perspectives. Int J Adv Intell Syst 12(1 & 2):1–13. http://www.iariajournals.org/intelligent_systems/

Schneider K (2019) What does competence mean? Psychology 10(14):1938–1958

Schuler D (2001) Cultivating society's civic intelligence: patters for a new 'World Brain.' Inf Commun Soc 4(2):157–181

Schuler D (2009) Communities, technology, and civic intelligence. In: Proceedings of the fourth international conference on communities and technologies (C&T '09). ACM, New York. https://www.researchgate.net/publication/221144093_Communities_technology_and_civic_intelligence

Schuler D (2010) Community networks and the evolution of civic intelligence. AI Soc 25(3):291–307. https://doi.org/10.1007/s00146-009-0260-z

Schuler D (2012) Civic intelligence and CSCW. In: Hercheui MD et al (eds) HCC10 2012, IFIP AICT 386. IFIP International Federation for Information Processing, pp 369–375

Schuler D (2014) Pieces of civic intelligence: towards a capacities framework. E-Learn Digit Media 11(5):518–529

Schuler D (2016) Smart cities + Smart citizens = Civic intelligence? In: Concilio G, Rizzo F (eds) Human smart cities: rethinking the interplay between design and planning. Springer, pp 41–60. https://doi.org/10.1007/978-3-319-33024-2_3

Schweitzer E (2017) On the road to becoming a smart citizen: defining, locating and supporting digital competences. Federal Institute for Research on Building, Urban Affairs and Spatial Development (BBSR), Bon. https://www.bbsr.bund.de/BBSR/EN/publications/AnalysenKompakt/Issues/ak-08-2017.html

Sharifi A (2019) A critical review of selected smart city assessment tools and indicator sets. J Clean Prod 233(1):1269–1283. https://doi.org/10.1016/j.jclepro.2019.06.172

Shelton T, Lodato T (2019) Actually existing smart citizens: expertise and (non)participation in the making of the smart city. City 23(1):35–52

Simonofski A, Asensio ES, De Smedt J, Snoeck M (2017) Citizen participation in smart cities: evaluation framework proposal. In: IEEE 19th conference on business informatics, Thessaloniki, Greece. https://doi.org/10.1109/CBI.2017.21

Simonofski A, Estefanía SA, Wautelet Y (2019) Citizen participation in the design of smart cities: methods and management framework. In: Visvizi A, Lytras M (eds) Smart cities: issues and challenges. Elsevier, pp 47–109

Star JR, Stylianides GJ (2013) Procedural and conceptual knowledge: exploring the gap between knowledge type and knowledge quality. Can J Sci Math Technol Educ 13(2):169–181. https://doi.org/10.1080/14926156.2013.784828

Steiner G, Risopoulos F, Mulej M (2013) Competences for citizen-driven innovation in crisis resolution. Syst Pract Action Res 26:571–577

Ten Dam G, Volman M (2004) Critical thinking as a citizenship competence: teaching strategies. Learn Instr 14:359–379

The Council of European Union (2018) Council Recommendation of 22 May 2018 on key competences for lifelong learning. Official J Eur Union C1289/1. https://eur-lex.europa.eu/legal-content/EN/TXT/PDF/?uri=CELEX:32018H0604(01)&rid=7

UNDP (United Nations Development Programme) (2010) Measuring capacity. https://www.undp.org/sites/g/files/zskgke326/files/publications/UNDP_Measuring_Capacity_July_2010.pdf

UNDG (United Nations Development Group) (2017) Capacity development: UNDAF companion guidance. https://unsdg.un.org/sites/default/files/UNDG-UNDAF-Companion-Pieces-8-Capacity-Development.pdf

UNECE (United Nations Economic Commission for Europe) (2020) People—Smart sustainable cities. United Nations, Geneva

United Nations (2016) The World's Cities in 2016. Data Booklet (ST/ESA/ SER.A/ 392). https://www.un.org/en/development/desa/population/publications/pdf/urbanization/the_worlds_cities_in_2016_data_booklet.pdf

Vuorikari R, Kluzer S, Punie Y (2022) DigComp 2.2: the digital competence framework for citizens—With new examples of knowledge, skills and attitudes. Publications Office of the European Union, Luxembourg. https://publications.jrc.ec.europa.eu/repository/handle/JRC128415

Wamsler C, Brink E (2014) Interfacing citizens' and institutions' practice and responsibilities for climate change adaptation. Urban Climate 7:64–91

Washburn D, Sindhu U, Balaouras S, Dines RA, Hayes NM, Nelson LE (2010) Helping CIOs understanding 'smart city' initiatives: defining the smart city, its drivers, and the role of the CIO. Forrester Research Inc., Cambridge, MA

Williamson B (2015) Educating the smart city: schooling smart citizens through computational urbanism. Big Data Soc July–December:1–13

Williamson B (2022). Smart citizen apprentices: digital urbanism and coding as techno-solutions to the city. In: Currie M, Knox J, McGregor C (eds) Data justice and the right to the city. Edinburgh University Press, Edinburgh, pp 165–184. https://edinburghuniversitypress.com/pub/media/ebooks/9781474492973.pdf

World Bank Group (2021) Demographic trends and urbanization. World Bank, Washington, DC. https://doi.org/10.1596/978-1-4648-1112-9

Xu H, Geng X (2019) People-centric service intelligence for smart cities. Smart Cities 2(2):135–152. https://doi.org/10.3390/smartcities2020010

Yang Y-TC (2012) Building virtual cities, inspiring intelligent citizens: digital games for developing students' problem solving and learning motivation. Comput Educ 59:365–377. https://doi.org/10.1016/j.compedu.2012.01.012

Yigitcanlar T, Kamruzzaman M, Buys L, Ioppolo G, Sabatini-Marques J, da Costa EM, Yun JJ (2018) Understanding 'smart cities': intertwining development drivers with desired outcomes in a multidimensional framework. Cities 81:145–160

Zhao F, Fashola OI, Olarewaju TI, Onwumere I (2021) Smart city research: a holistic and state-of-the-art literature review. Cities 119(103406):1–14

Data-Policy Interaction

The content of this paper, delineating the vital competencies essential for citizens' active engagement in shaping smart cities, closely aligns with the global policy agendas established by both the United Nations and the European Union. In congruence with the UN's 2030 Agenda for Sustainable Development, specifically Goal 11, which emphasizes sustainable cities and communities, the paper underscores the pivotal role of informed citizen participation in fostering inclusive and sustainable urban development. Additionally, it harmonizes with the principles outlined in The New European Consensus on Development, highlighting the necessity of citizen involvement in shaping urban policies. By emphasizing the importance of capacities for citizen participation in smart cities, the paper also aligns with the core competencies emphasized in the Reference Framework of Competences for Democratic Culture by the Council of Europe and the Council Recommendation on Key Competences for Lifelong Learning endorsed by the Council of the European Union.

Ivana Simić is a Full Professor at the Faculty of Economics, University of Niš, where she teaches courses in management and organization at undergraduate, graduate, and doctoral levels, including Fundamentals of Management, Management of Organizational Changes, and Management of Organizational Behavior. Her research interests also cover a wide range of topics, including entrepreneurship, intrapreneurship, digital transformation, personality, talent management, smart citizens, and algorithmic management. Throughout her academic career, she has authored or co-authored eight university textbooks and monographs and has published numerous scientific papers. She has also actively participated in 11 scientific research projects. She is a member of the Scientific Society of Serbian Economists.

Chapter 8
Key Determinants of the Quality of Towns and City Centres in the Function of More Resilient and Sustainable Tourism Development Management

Sonja Jovanović, Vesna Janković Milić, and Vinko Lepojević

Abstract Quality in the management of tourism development has become an imperative. In addition to the numerous benefits and results of growth that the tourism industry achieves in many countries, there are also risks and problems that destinations face. One of the key problems in maintaining the quality of destination is overcrowding, which is the subject of study in this paper. The pressure created by tourists causes serious environmental, social and economic consequences, as well as effects related to the perception of the destination by tourists. The first goal of this research is to examine the relationship between the number of tourists and the quality of towns and city centres in the most visited destinations in the world. A specific objective is to examine to what extent safety and security and pollution and environmental conditions have an impact on the quality of towns and city centres, expressed as accessible and pleasant to stay. Obviously, there is a need for balance between pleasurable tourist experience and residents' satisfaction.

Keywords Overtourism · Overcrowding · City centres · Quality of towns · Tourist satisfaction

8.1 Introduction

In the last few decades, tourism activities have recorded constant growth and become one of the fastest growing sectors in the world. It is often stated that tourism is an important factor that contributes to the development, both economic and social, of many destinations and countries. The data supporting this is that "travel and tourism

S. Jovanović (✉) · V. J. Milić · V. Lepojević
Faculty of Economics, University of Niš, Niš, Serbia
e-mail: sonja.jovanovic@eknfak.ni.ac.rs

J. J. Stanković et al. (eds.), *Regional and Urban Challenges and Opportunities for Data-Policy Interactions*, The Urban Book Series,
https://doi.org/10.1007/978-3-032-08592-4_8

(including its direct, indirect, and induced impacts) accounted for 1 in 5 new jobs created across the world during 2014–2019, and 10.3% of all jobs (334 million) and 10.4% of global GDP (US$ 10 trillion) in 2019. In 2022, the Travel and Tourism sector contributed 7.6% to global GDP" (WTTC 2023). International tourism arrivals in 2019 was 1,5 billion tourists (UNWTO 2020).

Since the 2000s, one of the key concepts of tourism development has been competitiveness. Some of the most significant works are the research by Crouch and Ritchie (1999) and then the creation of the Travel and Tourism Competitiveness Index by the World Economic Forum (2007). This concept encouraged destinations to compete and achieve better results compared to the competitors in terms of tourist numbers, economic results and market share. In the research from 2021, it is clearly stated that "level of competitiveness of destinations is an important contributing factor to potential overtourism situations" (Perles-Ribes et al. 2021). Authors (Perles-Ribes, J.F., Ramon-Rodriguez, A.B., Moreno-Izquirdo, L. and Such-Devesa, M.J.) point out in their research that in the case of Spain, as one of the most important tourist destinations in the world, "tourism competitiveness is a key predictor of overtourism".

The concept that has been attracting the attention of both theorists and tourism development policymakers for the last few years, as well as residents of cities and destinations with an excessive number of tourists is overtourism. The origin of this concept is connected with the appearance of publications on this topic since 2017. Overtourism mainly refers to the excessive number of tourists and the negative effects of tourism. Overcrowding represents an excessive number of tourists in one place or space, while overtourism is a broader concept that includes consequences of different character. The consequences are reflected in endangering the quality of life of the local population, the quality of water, air and the natural environment.

The subject of research in this paper is overtourism and the consequences it creates. Since this concept is very often linked to cities and the implications for local residents, the goal of the research is to examine the level of quality of centres in the most visited cities in the world. The special goal of the paper is to examine the level of influence of safety and security and pollution and environmental conditions on the quality of town and city centres in the most visited destinations. The information base consists of data from the World Economic Forum (WEF) on the Travel and Tourism Development Index (TTDI). The research is expected to indicate whether the most visited cities in the world are also the most threatened destinations in terms of overcrowding and accessibility.

8.2 Overtourism as a New Challenge of Sustainable Tourism Development

The concept of overtourism is also associated with carrying capacities. It is usual that in the literature "carrying capacities" refers to the ecological aspect or the spatial aspect of tourism. (Coccossis and Parpairis 2000) "Most destinations are managed based on a growth-paradigm, mainly valuing growth of visitor's numbers, without considering carrying capacity and other policy goals" (Peeters et al. 2018). When it comes to overtourism, one of the most important aspects is analysing social carrying capacity. Local residents are the ones who suffer the most from the consequences of overtourism. Thus, overtourism is most often linked to the negative consequences it creates for the local community. "Residents are very important part in the tourism sector, as their perceptions indicate tourism's outlook regarding sustainability" (Vagena 2021). The local residents' perception of tourists, their number and habits is very important in order to assess whether tourism in a city or destination can be characterized as overtourism. Thus, "overtourism describes destinations where hosts or guests, locals or visitors feel that there are too many visitors and that the quality of life in the area or the quality of the experience has deteriorated unacceptably" (Goodwin 2019, p.110.). In this way, it is indicated that overtourism is the opposite of sustainable tourism and the principles on which it is based.

To determine whether a destination is threatened by overtourism, the perception of tourists is also important. Namely, tourists can have different perceptions about overtourism. "People's tolerance of crowding is specific to the context" (Goodwin 2019, p. 112.). For tourists, crowds in shops or on the beach are an experience they expect in a destination. Therefore, determining whether a destination is threatened by the number of tourists or not depends on the attitude and experience of tourists. "The issues vary from place to place and it is important to focus on what matters locally and can be addressed locally" (Goodwin 2019, p.112.). Accordingly, "high level of crowding does not necessarily imply a lower-quality experience or a lower level of tourist satisfaction. Some crowding is experienced as positive and some is neutralized by coping behaviours, that reduce the disappointment and stress sometimes associated with crowding" (Peeters et al. 2018, p.32). Festivals or sports events can be mentioned as an example.

There are different definitions of overtourism. "Overtourism describes the situation in which the impact of tourism, at certain times and in certain locations, exceeds physical, ecological, social, economic, psychological and/or political capacity thresholds" (Peeters et al. 2018). This definition includes different forms of capacity that are unable to support the number of tourists in the location. The physical capacity of the destination is usually associated with the inability of the existing infrastructure to accommodate a large number of tourists (Marsiglio 2017). Ecological carrying capacity (Wenpeng et al. 2018; Kang and Xu 2010) refers to the condition and the level when the environment becomes degraded. Economic carrying capacity is a "threshold limit beyond which tourism growth becomes economically unacceptable" (Castelani et al. 2007, p. 366). After this limit, the costs for returning to the previous state become higher than the benefits that tourism brings.

In addition to physical, ecological, social and economic capacities, which are most often mentioned in the literature, the definition also includes psychological capacities (Zhiyong and Sheng 2009) which refer to "capacity of people (residents and/or other visitors) to emotionally cope with crowding effects" (Peeters et al. 2018, p.22). The political capacity is also highlighted, which is linked to the local government's ability to manage the consequences that overtourism creates for the local population. If any of the mentioned capacities are exceeded, there are consequences of overtourism.

Overtourism can be defined as "the impact of tourism on destination, or parts thereof, that excessively influences perceived quality of life of citizens and/or quality of visitors experiences in a negative way" (UNWTO 2018, p.4, according to universities collaborating in the project). This definition of overtourism especially emphasizes urban tourism and towns and cities, which are becoming increasingly popular destinations. Social sustainability in cities becomes an important issue, as well as social carrying capacities.

When Airbnb started its "Live like a local" campaign in 2016, it is often mentioned in the literature as the moment when the pressure of tourists on the local population in cities definitely became even greater. The New York Times published an article saying that Airbnb "began offering Guidebooks, a feature that lets those who rent out their homes on Airbnb share information about their neighbourhood's best restaurants, bars and attractions, including local spots that might not be found on travel websites or guides, all in the Airbnb app" (New York Times 2016).In this way, an ambience was created where "inhabitants and visitors compete for the access to exactly the same places that are precious assets for both these groups" (Żemła 2020, p.13).It is possible for tourists to be as close as possible to local residents and to stay in their neighbourhood. Apart from that, "outside of regular hotel capacities, it is now possible to rent a private apartment during a visit, probably owned by residents who have left because of the crowds, prices and lack of city function" (Hristov and Danilović Hristić 2021, p. 59).

The authors of Perkumenie and Pranskuniene define overtourism in a wider area, that is, they look at the consequences of an excessive number of tourists in the entire region. According to these authors, "overtourism is characterized by an excessive number of visitors, which affects the quality of the region" (Perkumienė and Pranskūnienė 2019).

The *causes of overtoruism* are numerous. One of the reasons is the exclusive focus of tourism development strategies and programmes on the number of tourists and the economic benefits it brings. Overtourism is spread to all destinations. Tourists' interests in getting to know other, more distant destinations, as well as the life of people in those destinations, lead to overtourism becoming present on all continents. An increase in international tourist arrivals to a country can also be added here. Another cause of overtourism is more accessible air transportation. The expansion of the services of low-cost airlines and the lower prices at which they offer their flights are becoming more accessible to more people. "With respect to transport, the technological evolution makes it easier to purchase cheaper, faster and safer trips both in the airline and cruises" (Garcia-Hernandez et al. 2019, p.8). One important pattern that contributes to overtourism is the expansion of the use of digital platforms, such as P2P platform activities. With the help of these platforms, it is easy to find adequate, affordable accommodation, which is located in buildings or in the neighbourhood where the local population lives. These platforms contribute to the territorial concentration of tourists and their number in a certain destination or location. In addition to spatial, temporal concentration of tourists is also the cause of overtourism. The presence of seasonality in tourism creates a concentration of tourists within one period of the year at a specific location. A significant contribution to the overtourism is made by "the dissemination of online information with recommendations (Tripadvisor.com), social networks (Facebook, Instagram and similar with billions of followers), the influence of media and massive publicity, easy online booking (companies such as Booking.com, Expedia.com, Trivago.com, Agoda.com, Kayak.com, Otel.com, etc.)" (Hristov and Danilović Hristić 2021, p.62).

Indicators that show that there is overtourism in a city or destination, according to Peeters et al. (2018) are as follows (Peeters et al. 2018, p. 16):

- tourism density (bed-nights per km^2) and intensity (bed-nights per resident);
- the share of Airbnb bed capacity of the combined Airbnb and booking.com bad capacity;
- the share of tourism in regional GDP;
- air travel intensity (arrivals by air divided by number of residents); and
- closeness to airport, cruise port and UNESCO World Heritage Sites.

Monitoring indicators such as tourism density or tourism intensity over time is very important in order to assess the trend and make assumptions about the future growth of tourism activity. In this way, it is possible to prevent overtourism in destinations by taking adequate measures. "Avoiding overtourism requires custom-made policies in cooperation between destinations' stakeholders and policymakers" (Vagena 2021, p.4).

The most common *consequences of overtourism* are degraded tourist experience, overloaded infrastructure, damage to nature, threats to culture and heritage. One of the consequences is the loss of authenticity of spaces or historical buildings. This will further result, in a longer period of time, in reducing the attractiveness of the destination and will certainly affect its demand and the future number of tourists.

The *challenge* that overtourism creates is to provide a positive tourist experience. That is, the quality of the tourists' experience is very important in order for the destination to survive and its development to be sustainable.

There are a greater number of measures that are proposed to mitigate the consequences of overtourism or to create an environment in which carrying capacities will not be exceeded. "The most frequent measures taken by destination management organizations and local governments to soften the negative effects of overtourism are related to spreading visitors in time and space" (Peeters et al. 2018, p. 17). These measures can include extending the attractiveness season of a destination, improving infrastructure capacity, marketing activities for the promotion of attractions located outside the city centre or near the destination.

"Overtourism is often associated with cities and urban tourism but is also a common phenomenon in rural, coasts and islands' settings, and at natural and cultural heritage sites and large attractions" (Peeters et al. 2018, p. 24). In literature, there are two main attitudes towards overtourism. "The first attitude, the wide approach, perceives overtourism as a phenomenon that can be seen in many different places" (Żemła 2020). This refers to coastal, islands and rural heritage sites. Another approach towards overtourism refers to towns and cities. In the continuation of this work, the focus is on cities that face the challenges of overtourism. The emphasis is on the consequences of an excessive number of tourists, but also on the reasons that lead to these problems.

8.3 Overtourism in Towns and Cities

"Over half of the world's population lives in urban areas and it is estimated that, by 2050, this proportion will reach 70%. In addition, the growing number of urban tourists increases the use of natural resources, causes socio-cultural impact and exerts pressure on infrastructure, mobility and other facilities" (UNWTO 2018, p. 3). Cities are becoming increasingly attractive destinations for tourists. "The increasing demand for mobility, leisure and unique experience has been examined as a key factor in the growth of tourism, causing pressure on locations and dependence on tourism" (Hristov and Danilović Hristić 2021, p.45). Urban tourism allows tourists to satisfy a large number of demands and expectations when travelling.

"The growth of urban tourism creates important challenges to ensure sustainable growth and practices that minimize any adverse effects that the development of tourism may have in terms of the use of natural resources, socio-cultural impact, pressure on infrastructure, mobility and congestion management" (UNWTO 2018, p.4). Local authorities get additional tasks that involve managing the increased tourists demand and the capacities that the destination has.

"In a very short time, the term overtourism has become to be used to describe the negative effect of tourism, and has been applied to the problem of excessive number of tourists in many cities" (Kruczek 2019). Attention is focused on the consequences of an excessive number of tourists on the lives of local residents, waste, pollution, traffic jams, visitor behaviour, rising costs of living for local residents, increased water and energy consumption and more. It is clear that overtourism is not exclusively associated with a large and growing number of tourists, but with the consequences, it brings with it and the carrying capacity of the destination to support the pressure of a large number of tourists.

It is important for every destination to attract tourists and to realize numerous benefits based on their number. However, the number of tourists in the destination becomes an important problem due to the phenomenon of overtourism. Managing the number of tourists in destinations, such as cities, is very complex and there are not many options for solving it. The measures are mainly related to marketing activities, with the help of which it is expected to redirect the demand for certain destinations. Some of the proposed measures for regulating tourist arrivals are (Goodwin 2019, p.114):

- demarketing to deter particular market segments or reduce demand pressure,
- use marketing to change the destination image and attract more compatible market segments,
- use marketing and regulation to change tourist behaviour,

- tourist taxes.

Research related to overtourism was carried out in many European cities, known as attractive tourist destinations. In cities, there is a problem of regulating the number of tourist arrivals. "For example, in Barcelona overtourism is to a significant extent a consequence of the ease of arriving in the city by plane, sea, rail or road. The city authorities are not able to manage arrivals through the port, airport or railway station" (Goodwin 2019, p.112). Possibilities for this regulation still exist, but research, availability of data and prediction of effects are necessary. "It is not possible, nor desirable, to describe precise and exact policy measures because there is scarce empirical evidence to found such measures on" (Peeters et al. 2018, p.15). Also, it is not possible to apply the same pattern of measures in every city or destination. However, some recognized measures to manage overtourism in cities, which can give significant results, can be:

- regulating public transport and directing tourists to other locations;
- spatial planning that will regulate the location of stores;
- restrictions related to the approach of cruise ships.

Local government, tourist organizations, destination management organizations, according to UNWTO, can apply other measures (UNWTO 2018, p.7):

- improve infrastructure and facilities in the city;
- communicate with and involve local residents and local business in tourism planning;
- communicate better with visitors on how to behave in the city;
- distribute visitors better over the year;
- create city experiences where residents and visitors can meet and integrate.

The *consequences of overtourism* refer to environmental, economic and social impact. However, the impact of overtourism is different in relation to the type of destination. Thus, based on research that included 41 destinations (Peeters et al. 2018), it was found that "social impacts prevail in urban destinations, environmental impacts in rural, while all three impact categories are relevant in coastal and islands and heritage and attractions". Therefore, residents of cities that are attractive tourist destinations and attract a large number of tourists show their dissatisfaction and intolerance towards tourists. The reason is the inability of local residents to carry out their usual personal activities every day (going to the store, going to the market, walking, visiting cafes and restaurants), but also excessive noise, waste, crowds. Other negative consequences of excessive tourism in cities refer to the increase in prices in shops, restaurants and even in public transport, then "real estate speculation, deterioration of local residents' identification with the place, loss of a destination's authentic character, congestion of transport infrastructure" (Żemła 2020, according to Benner 2019). Preservation of local traditions and values is important so that the inhabitants of tourism cities do not move out in search of places where they will have an adopted lifestyle. "There is no doubt that cities depend on tourism because it creates significant income, but balance should be achieved, keeping the traditional lifestyle in place" (Hristov and Danilović Hristić 2021, p. 61).

In addition to exclusively negative consequences, overtourism can be seen in the light of positive changes and contributions. The results of the research (UNWTO 2018, p. 7) which was conducted in eight European cities (Barcelona, Berlin, Amsterdam, Copenhagen, Lisbon, Munich, Salzburg and Tallinn) showed "that the most relevant positive impacts from tourism as perceived by residents are: greater international atmosphere (different cultures in the city); more events; a more positive image; protection of historical parts of the city; and restorations of traditional architecture".

In this way, it was pointed out that in some cities there is a greater or lesser degree of tolerance on the part of the local population towards overtourism. The explanation in their research is given by the authors (Perles-Ribes, J.F., Ramon-Rodriguez, A.B., Moreno-Izquirdo, L. and Such-Devesa, M.J.) who indicate that "tourism's share of Gross Domestic Product (GDP) among the drivers of overtourism". Namely, "in a city with a diversified economy, where the main source of the wealth of its residents is an activity other than tourism (industry, financial services, etc.), there would be a low level of tolerance of the nuisance generated by tourists" (Perles-Ribes et al. 2021). In this situation, "the inhabitants who do not see any economic benefit (and also benefits of other kinds) are not willing to limit their needs and adjust their way of living to satisfy the visitor's needs" (Żemła 2020, p. 8). However, in cities where tourism represents the basic economic activity and the main source of income for the population, the tolerance of the local population will be at a much higher level in relation to overtourism.

Research related to overtourism in cities was presented by the authors Perez Garrido et al. (2022) on the example of Budapest. They determined that "the centre of Budapest has been strongly affected by the presence of tourists, while districts physically far from the centre have been affected". They indicated the consideration of the heterogeneity of the city and the spatial direction of tourists to other parts of the city. This is the task of policymakers and destination management organizations, who should recognize the problem through marketing activities, direct public transport, redirect tourists and ensure a sustainable number of tourists in the entire city.

"City administrators and destination managers must acknowledge that there are definite limits to growth. Prioritizing the welfare of local residents above the needs of the global tourism supply chain is vital" (Milano et al. 2018). Appreciating the destination's capacities is key to ensuring the further development of tourism in cities.

Different solutions are considered as suitable to deal with the growing number of tourists. Research (Garcia-Hernandez et al. 2019) conducted in four Spanish destinations (Donostia-San Sebastian, Madrid, Malaga and Valencia) aimed to examine the level of implementation of smart solutions in cities where overtourism is present. Results showed "that issues related to overtourism in four Spanish destinations still have little impact on urban tourism planning in general and on smart city and smart destination strategies in particular" (Garcia-Hernandez et al. 2019, p. 5). Application of smart and technological solutions, according to these authors, is "weak and there is a gap between smart destination approaches the level of implementation of smart solutions and their operability in relation to the management of tourist overload".

Pointing out the problems caused by overtourism requires finding sustainable models for the future development of tourism in cities. The solutions that some cities have taken in dealing with an unsustainable number of tourists and the consequences that their presence creates can imply "demotions" of tourists. "It is not desirable or possible to avoid tourism, but for those suffering from exaggerated levels of visitor flow, some kind of strategy or plan for moderation should be implemented" (Hristov and Danilović Hristić 2021, p.64).

8.4 Analysis and Discussion

The research in this paper focuses on the most visited cities in the world. The goal of the research is to examine the relationship between the number of tourists and the quality of towns and city centres in the most visited cities in the world. A specific objective is to examine to what extent safety and security and pollution and environmental conditions have an impact on the quality of towns and city centres, expressed as accessible and pleasant to stay.

The research will analyse data for 2019, bearing in mind that due to the COVID 19 pandemic, data for 2020 and 2021 are not relevant for the presentation of the objective attendance of cities.

(1) The goal of the first part of the analysis is to examine whether the most visited cities in the world are located in countries that are rated as the best according to the quality of towns and city centres.

Table 8.1 shows the ten most visited cities in 2019 in terms of international overnight visitors. There is a significant difference (more than 12 million visitors) between the most visited city—Hong Kong, with 26.7 million visitors, compared to Kuala Lumpur, which is in 10th place with 14.1 million visitors. Even two cities in China are among the ten most visited in the world—Hong Kong and Macau.

In further analysis, the best-rated countries in the world according to quality of city centres will be shown. The information base for this analysis is data from the Travel and Tourism Development Index 2022 (WEF 2022). "This index serves as a strategic benchmarking tool for policymakers, companies and complementary sectors to advance the future development of the Travel and Tourism (T&T) sector by providing unique insights into the strengths and development areas of each country/economy to enhance the realization of sector potential and growth" (WEF 2022). The index is comprised of five subindexes, 17 pillars, and 112 individual indicators. Within the subindex—*T&T Sustainability*, there are three pillars—Environmental Sustainability, Socioeconomic Resilience and Conditions and T&T Demand Pressure and Impact. One of indicators within pillar—*T&T Demand Pressure and Impact* is—*Quality of town and city centres.* "Here are specific efforts that can be made to manage T&T to prevent overcrowding and efficiently use a destination's carrying capacity"

Table 8.1 World's most visited cities in 2019 (in million)

Rank	City	Country	Number of visitors	Rank	City	Country	Number of visitors
1	Honk Kong	China	26.70	6	Paris	France	19.10
2	Bangkok	Thailand	25.80	7	Dubai	United Arab Emirates	16.30
3	Macau	China	20.60	8	Delhi	India	15.20
4	Singapore	Singapore	19.80	9	Istanbul	Turkey	14.70
5	London	United Kingdom	19.60	10	Kuala Lumpur	Malaysia	14.10

Source Govisity (2022) The 50 Most Visited Cities in the World and their Top Tourist Attractions, according to Mastercard's Global Destinations Cities Index

(WEF 2022, p.30). The evaluation and ranking of countries according to the quality of city centres was carried out on the basis of a questionnaire and answering the question: "In your country, how would you characterize towns and city centres?" Answering this question is ranked on a scale from 1 to 7, where 1 indicates that the centres are overcrowded and/or accessible only to some members of the public, while 7 indicates that the centres are widely accessible and pleasant to be in (WEF 2022).

Table 8.2 shows the ten best-rated countries in the world according to the quality of town and city centres.

It is noted that not a single country received the maximum score in terms of quality of town and city centres. This means that there is room for improving the quality of cities and towns, reducing overcrowding, spatial dispersion of tourists through the promotion of attractions in wider areas in cities. Table 8.3 shows that only two countries with the best-rated quality of town and city centres—Singapore and the United Arab Emirates—are also the countries with the most visited cities in the world. Table 8.3 shows the ranking and scores of the quality of town and city centres of the countries with the most visited cities in the world in 2019.

Table 8.2 Quality of town and city centres, 2019

Rank	Country	Score	Rank	Country	Score
1	Finland	6.4	6	United States	5.9
2	Austria	6.1	7	Luxemburg	5.9
3	Singapore	6.1	8	Lithuania	5.7
4	Switzerland	6.0	9	Korea	5.7
5	Iceland	6.0	10	United Arab Emirates	5.6

Source World Economic Forum (2022) Travel & Tourism Development Index 2021: Rebuilding for a Sustainable and Resilient Future, https://www3.weforum.org/docs/WEF_Travel_Tourism_Development_2021.pdf

Table 8.3 Score and rang of quality of town and city centres in countries with most visited cities in the world, 2019

Rank	Country	Rank in the world list	Score	Rank	Country	Rank in the world list	Score
1	China	12	5.6	6	France	41	4.9
2	Thailand	78	4.0	7	United Arab Emirates	10	5.6
3	China	12	5.6	8	India	59	4.4
4	Singapore	3	6.1	9	Turkey	80	3.9
5	United Kingdom	47	4.8	10	Malesia	37	5.0

Source World Economic Forum (2022) Travel & Tourism Development Index 2021: Rebuilding for a Sustainable and Resilient Future, https://www3.weforum.org/docs/WEF_Travel_Tourism_Development_2021.pdf

Table 8.3 shows that, except for the United Arab Emirates and Singapore, all other countries with the most visited cities in the world have very low ranks and ratings of the quality of towns and city centres. In relation to the 117 countries included in the Travel and Tourism Development Index Report, Turkey is in 80th place, Thailand in 78th place, India in 59th place, while Malaysia is in 37th position, France is in 41st place, United Kingdom is in 47th place and China is in 12th position. The range of scores is wide (from 3.9 to 6.1). This indicates that a significantly larger number of countries that have cities with the largest number of tourists are also countries in which the quality of town and city centres is rated very poorly. For these countries and their tourist destinations, it is necessary to consider carrying capacities, strategies and models of sustainability.

(2) The second part of the analysis aims to examine to what extent safety and security and pollution and environmental conditions have an impact on the quality of towns and city centres. The information basis for the research in this part of the work is data from the Report of the World Economic Forum on the Travel and Tourism Development Index (WEF 2022). The selection of indicators for analysis was based on the assessment that exposes locals and tourists to pollution and security risk are very important for attracting tourists in a country.

Safety and security is pillar within TTDI Subindex—*Enabling Environment* and includes six components—Business costs of crime and violence, reliability of police services, safety walking alone at night, homicide rate, Global Terrorism Index and organized violence. "This pillar measures the extent to which a country exposes locals, tourists and businesses to security risks" (WEF, 2020) Value of this pillar is in range of 1 to 7, where 1 means high exposes tourists to security risk, and 7 means low security risk. *Pollution and environmental conditions* is a subpillar within the pillar—*Environmental Sustainability* and consists of components showing particulate matter concentration, baseline water stress, waste water treatment and clean ocean water.

Value of this subpillar is also in range of 1 to 7, where 1 means high air and water pollution and 7 means clean environment.

Table 8.4 shows the score and rank of safety and security in countries with most visited cities in the world. An extremely wide range of rankings is observed in the world list, which includes 117 countries. Singapore is the first country in the world for the safety of tourists. This country is the safest and best rated in relation to all countries according to the component—*safety walking alone at night*. The United Arab Emirates ranks third in the world according to safety and security, while the other seven analysed countries are ranked and rated very low. Thailand, India and Turkey are at the very bottom of the world list, at 88th, 86th and 98th place, respectively. The range of scores is from 4.4 to 6.7, which represents wide heterogeneity between countries.

The analysed group of countries with most visited cities in the world, in Table 8.5, is ranked very poorly according to the degree of pollution and environmental conditions. The best ranked and rated country is the United Kingdom (8th place), while all other countries are on the world ranking list from 25th place (France) to 117th place (India). India is the worst ranked country in the world according to pollution and environmental conditions. The range of grades is from 2.7 to 5.7, which indicates that these countries are the most heterogeneous according to the parameters of pollution and environmental conditions. Six of the ten analysed countries are in the second half of the world list according to this parameter.

In further analysis within this part of the work, the 50 most visited cities around the world (Govisity 2022) in 2019 were included. The most visited cities in the world are located in 27 countries. China, India and the USA have the largest number of most visited cities in the world. The three most visited cities are located in Thailand, Saudi Arabia and Italy, while Turkey, Spain, Malaysia, Japan and Vietnam each have the two most visited cities in the world (Table 8.6).

Table 8.4 Score and rang of *safety and security* in countries with most visited cities in the world, 2019

Rank	Country	Rank in the world list	Score	Rank	Country	Rank in the world list	Score
1	China	52	5.5	6	France	55	5.5
2	Thailand	88	4.8	7	United Arab Emirates	3	6.5
3	China	52	5.5	8	India	86	4.9
4	Singapore	1	6.7	9	Turkey	98	4.4
5	United Kingdom	44	5.6	10	Malesia	51	5.5

Source World Economic Forum (2022) Travel & Tourism Development Index 2021: Rebuilding for a Sustainable and Resilient Future, https://www3.weforum.org/docs/WEF_Travel_Tourism_Development_2021.pdf

Table 8.5 Score and rang of *pollution and environmental conditions* in countries with most visited cities in the world, 2019

Rank	Country	Rank in the world list	Score	Rank	Country	Rank in the world list	Score
1	China	109	3.3	6	France	25	5.0
2	Thailand	98	3.5	7	United Arab Emirates	87	3.7
3	China	109	3.3	8	India	117	2.7
4	Singapore	66	4.1	9	Turkey	56	4.2
5	United Kingdom	8	5.7	10	Malesia	78	3.9

Source World Economic Forum (2022) Travel & Tourism Development Index 2021: Rebuilding for a Sustainable and Resilient Future, https://www3.weforum.org/docs/WEF_Travel_Tourism_Development_2021.pdf

Table 8.6 Countries with the most visited cities in the world in 2019

	Country	Number of cities		Country	Number of cities
1	China	5	15	France	1
2	India	5	16	Singapore	1
3	USA	5	17	Czech Republic	1
4	Thailand	3	18	South Korea	1
5	Saudi Arabia	3	19	Netherlands	1
6	Italy	3	20	Indonesia	1
7	Turkey	2	21	Russia	1
8	Spain	2	22	United Arab Emirates	1
9	Malaysia	2	23	Egypt	1
10	Japan	2	24	Austria	1
11	Vietnam	2	25	Greece	1
12	United Kingdom	1	26	Germany	1
13	Taiwan	1	27	Ireland	1
14	Mexico	1			

Source Govisity (2022) The 50 Most Visited Cities in the World and their Top Tourist Attractions, https://govisity.com/most-visited-cities-in-the-world/

Data on the number of tourists in 50 cities were put into relation with the ratings of Quality of town and city centres, safety and security and pollution and environmental conditions in their countries.

Quality of town and city centres was observed as a dependent variable, while the independent variables are—safety and security and pollution and environmental conditions (Table 8.7).

Table 8.7 Variables included in the model

Variable	Variable type
Quality of town and city centres	Dependent
Safety and security	Predictor
Pollution and environmental conditions	Predictor

Based on the data in Table 8.8, we find that the highest average value and the lowest standard deviation are for safety and security, while the lowest average value and the highest standard deviation are for pollution and environmental conditions.

Based on the data from Table 8.9, it can be seen that there is a strong direct relationship between quality of town and city centres and safety and security, while there is a moderately strong direct correlation between the dependent variable and pollution and environmental conditions. These parameters indicate that the improvement of safety and security in cities, as well as environmental conditions, is accompanied by a higher level of quality in city centres in terms of accessibility and a pleasant stay for tourists and the local population.

The regression model that observes the impact of safety and security and pollution and environmental conditions on the quality of town and city centres explains 50.7% of the total variability of the dependent variable (Table 8.10).

According to this model, there is a direct relationship between the dependent variable (quality of town and city centres) and two independent variables (safety and security and pollution and environmental conditions). The model shows that if the value of one of these two independent variables increases, it will result in an increase in the dependent variable. The variable safety and security contributes

Table 8.8 Descriptive statistics

Variable	Mean	Std. deviation
Quality of town and city centres	4.9653	0.63132
Safety and security	5.4367	0.59746
Pollution and environmental conditions	4.2122	0.96450

Table 8.9 Correlation matrix

	Quality of town and city centres	Safety and security	Pollution and environmental conditions
Quality of town and city centres	1.000	0.693[a]	0.399[a]
Safety and security		1.000	0.358[a]
Pollution and environmental conditions			1.000

[a]. Correlation is significant at the 0.01 level

Table 8.10 Regression parameters

	Beta		t	Sig
	Unstandardized	Standardized		
(Constant)	0.861		1.427	0.160
Safety and security	0.667	0.631	5.692	0.000
Pollution and environmental conditions	0.113	0.173	1.562	0.125
R square	0.507[a]			

[a]. Dependent Variable: Quality of town and city centres

significantly more to the explanation of the variability of the dependent variable in relation to pollution and environmental conditions. Improving safety and security in a country greatly affects the quality of city centres. Reducing pollution and improving environmental conditions also contribute to the quality of city centres, but to a lesser extent in relation to safety and security.

8.5 Conclusion

Increase in international tourist arrivals causes serious environmental, social and economic effects, as well as effects related to the perception of the destination by tourists. The negative consequences created by an excessive number of tourists in a destination are described by the concept of overtourism. There are many impacts of overtourism: overloaded infrastructure, damage to nature, threats to culture and heritage, loss of authenticity of spaces or historical buildings and degraded tourist experience. Cities are becoming increasingly attractive destinations for tourists. Managing the number of tourists in destinations, such as cities, is very complex and there are not many options for solving it. The most frequently proposed measure is related to marketing activities and directing tourist demand towards other attractions located in the wider city area. In addition to the territorial one, temporal dispersion of tourists is proposed through the activities of expanding the tourist season and redistributing the number of tourists throughout the year.

The research in this paper focuses on the most visited cities in the world. The analysis in the paper showed that the cities with the largest number of tourists in the world (Hong Kong, Bangkok, Macau, Singapore, London, Paris, Dubai, Delhi, Istanbul, Kuala Lumpur) are located in the countries (China, Thailand, Singapore, United Kingdom, France, United Arab Emirates, India, Turkey, Malaysia) which are not, according to WEF, rated as countries with the best quality of city centres. Namely, only two countries with the best-rated quality of town and city centres, Singapore and the United Arab Emirates, are also the countries with the most visited cities in the world. All other countries with the most visited cities in the world have very low ranks and ratings of the quality of towns and city centres. In relation to the 117 countries included in the Travel and Tourism Development Index (WEF), Turkey is in 80th place, Thailand in 78th place and India in 59th place. This indicates that a significantly larger number of countries that have cities with the largest number of tourists are also countries in which the quality of town and city centres is rated very poorly. For these countries and their tourist destinations, it is necessary to consider carrying capacities, strategies and models of sustainability. It is important for these cities to better manage public traffic, redirect tourists to other parts of the city, and place greater emphasis on the quality of life of the residents of the city centres, that is, on the social component of urban living.

Further research in this paper was related to determining the factors that contribute to the greatest extent to the quality of city centres in the most visited cities in the world. Singapore (the fourth most visited country in the world in 2019) is the first country in the world for the safety of tourists. The analysed group of countries with most visited cities in the world is ranked very poorly according to the degree of pollution and environmental conditions. The best ranked and rated country is the United Kingdom (8th place), while all other countries are on the world ranking list from 25th place (France) to 117th place (India). India is the worst ranked country in the world according to pollution and environmental conditions, while on the other hand, the city of Delhi in 2019 had over 15 million visitors.

The research in this paper has clearly shown that the most visited cities in the world are located in countries (except Singapore and the United Arab Emirates) where the quality of cities is not rated at the highest level. This indicates that these cities are threatened by an excessive number of tourists, that is, that there is overtourism. The analysis in the paper showed that the improvement of safety and security in cities, as well as environmental conditions, is accompanied by a higher level of quality of city centres in terms of accessibility and a pleasant stay for tourists and the local population. Reducing pollution and improving environmental conditions contribute to the quality of city centres to a lesser extent in relation to safety and security.

Funding This study was supported by the European Union's Horizon Europe research and innovation programme WIDERA-2021-ACCESS-02 under grant number 101059994—UR-Data.

References

Benner M (2019) From overtourism to sustainability: a research agenda for qualitative tourism development in the Adriatic. MPRA Paper 92213, University Library of Munich, Germany

Castelani V, Sala S, Pitea D (2007) A new method of tourism carrying capacity assessment. In: Ecosystems and sustainable development (eds) Tiezzi E, Marques JC, Brebbia CA, Jorgenson SE, WIT Press, UK., pp 365–374

Coccossis H, Parpairis A (2000) Tourism and the environment: some observations on the concept of carrying capacity. In: Briassoulis H, van der Straaten J (eds) Tourism and the Environment. Environment and Assessment, vol 6. Springer

Crouch GI, Ritchie JRB (1999) Tourism, competitiveness, and societal prosperity. J Bus Res 44(3):137–152

Garcia-Hernandez M, Ivars-Baidal J, Mendoza de Miguel S (2019) Overtourism in urban destinations: the myth of smart solutions. Bol AGE 83:2830

Goodwin H (2019) Overtourism: causes, symptoms and treatment, Tourismus Wissen

Govisity (2022) The 50 most visited cities in the world and their top tourist attractions. https://govisity.com/most-visited-cities-in-the-world/. Last accessed 18 Apr 2023

Hristov M, Danilović Hristić N (2021) Impact of overtourism on urban life. Spatium (45):59–66

Kang P, Xu L (2010) The urban ecological regulation based on ecological carrying capacity. Procedia Environ Sci 2:1692–1700

Kruczek Z (2019) Overtourism—around the definition. https://www.researchgate.net/publication/333186191_OVERTOURISM_-_AROUND_THE_DEFINITION#fullTextFileContent. Last accessed 18 Apr 2023

Marsiglio S (2017) On the carrying capacity and the optimal number of visitors in tourism destinations. Tour Econ 23(3):632–646

Milano C, Cheer JM, Novelli M (2018) Overtourism is becoming a major issue for cities across the globe. World Economic Forum. https://www.weforum.org/agenda/2018/07/overtourism-a-growing-global-problem. Last accessed 25 Apr 2023

New York Times (2016) Airbnb Wants Travelers to 'Live Like a Local' With Its App, by Katie Benner

Peeters P, Gössling S, Klijs J, Milano C, Novelli M, Dijkmans C, Eijgelaar E, Hartman S, Heslinga J, Isaac R, Mitas O, Moretti S, Nawijn J, Papp B, Postma A (2018) Research for TRAN committee—overtourism: impact and possible policy responses, European Parliament, Policy Department for Structural and Cohesion Policies, Brussels

Perez Garrido B, Sebrek SS, Semenova V, Bal D, Michalko G (2022) Addressing the phenomenon of overtourism in Budapest from multiple angles using unconventional methodologies and dana. Sustainability 14:2268

Perkumienė D, Pranskūnienė R (2019) Overtourism: between the right to travel and residents' rights, Sustainability 11(7):2138. https://doi.org/10.3390/su11072138. Last accessed 18 Apr 2023

Perles-Ribes JF, Ramon-Rodriguez AB, Moreno-Izquirdo L, Such-Devesa MJ (2021) Competitiveness and overtourism: a proposal for an early warning system in Spanish urban destinations. Eur J Tour Res 27:2707

UNWTO (2020) World tourism barometer and statistical annex. https://doi.org/10.18111/wtobarometereng.2020.18.1.1. Last accessed 4 Apr 2023

Vagena A (2021) Overtourism: definition and impact, academia letters, Article 1207

Wenpeng D, Huimin Y, Yanzhao Y, Fang L (2018) Evaluation methods and research trends for ecological carrying capacity. J Resour Ecol 9(2):115–124

World Economic Forum (2007) Travel and tourism competitiveness report: paving the way for a more sustainable and inclusive future. https://www3.weforum.org/docs/WEF_TTCR_2017_web_0401.pdf. Last accessed 25 Apr 2023

World Economic Forum (2022) Travel and tourism development index 2021: rebuilding for a sustainable and resilient future. https://www3.weforum.org/docs/WEF_Travel_Tourism_Development_2021.pdf. Last accessed 2 Apr 2023

World Tourism Organization (UNWTO) Centre of Expertise Leisure, Tourism & Hospitality; NHTV Breda University of Applied Sciences; and NHL Stenden University of Applied Sciences (2018) Overtourism'?—Understanding and managing urban tourism growth beyond perceptions, executive summary, UNWTO, Madrid, https://doi.org/10.18111/9789284420070. Last accessed 25 Apr 2023

WTTC (2023) Economic Impact Reports. https://wttc.org/research/economic-impact. Last accessed 25 Apr 2023

Żemła M (2020) Reasons and consequences of overtourism in contemporary cities—knowledge gaps and future research. Sustainability 12:1729. https://doi.org/10.3390/su12051729. Last accessed 25 Apr 2023

Zhiyong F, Sheng Z (2009) Research on psychological carrying capacity of tourism destination. Chin J Popul Resour Dev 7(1):47–50

Smart Growth in Cities

Chapter 9
On Foot Through Town? The Conundrum of Pedestrian Accessibility in Niš, Serbia

Claudia van der Laag, Wendy Tan, Chiara Garau, and Beniamino Murgante

Abstract This chapter focuses on potential and realized accessibility and how inclusive the pedestrian experience is within the city of Niš, Serbia. This would require an assessment of the city's accessibility potential, and designing and planning urban spaces to fit with the users' experience and perceptions. To date, little is known about the potential and realized accessibility for pedestrians in socialist-influenced cities of the Western Balkans. This chapter therefore aims to understand the relationship between potential and realized pedestrian accessibility for this city which shows how residents rely on the car for their basic needs. For this case study, we applied a combined quantitative and qualitative approach of (1) computational street network analysis of street centrality and accessibility using space syntax, (2) semi-structured interviews to capture users' understanding perceptions of inclusivity and (3) visual analysis of the built environment. Our findings indicated that inclusive pedestrian accessibility was not fully realized due to unsupportive urban street design. Niš's urban fabric indicated a high potential for pedestrian accessibility in the historically grown structures and less potential in socialist and modernist neighbourhoods. The socialist city planning impact on Niš historic city structure was evident through urban design and architecture that neglected pedestrians. The socialist neighbourhoods accommodated car-based planning and therefore generated a non-inclusive environment on all scales with the exception of key public spaces. This study provides insights into the city of Niš urban fabric's functionality and how urban design and

---

C. van der Laag (✉)
Department of Built Environment, Oslo Metropolitan University, Oslo, Norway
e-mail: claudia.vanderlaag@oslomet.no

W. Tan
Wageningen University Research, Wageningen, The Netherlands

Western Norway University, Bergen, Norway

C. Garau
DICAAR, University of Cagliari, Cagliari, Italy

B. Murgante
School of Engineering, University of Basilicata, Potenza, Italy

J. J. Stanković et al. (eds.), *Regional and Urban Challenges and Opportunities for Data-Policy Interactions*, The Urban Book Series,
https://doi.org/10.1007/978-3-032-08592-4_9

planning for inclusive pedestrian facilities can contribute to a safe, inclusive urban environment for pedestrians and vulnerable groups.

Keywords Pedestrian accessibility · Space syntax · People's perception · Urban street design

9.1 Introduction

The concept of inclusivity for pedestrian accessibility involves designing and planning urban spaces in a way that accommodates people of all abilities and backgrounds. The aim is to create safe and welcoming spaces that are accessible to everyone. This shift from functional accessibility to inclusive accessibility is motivated by ideas such as universal design (Bringa 2018), access-for-all (Clarke 2004; Venter et al. 2019) and reducing mobility inequality (Hidayati et al. 2021a, b). Accessibility is a crucial factor in achieving inclusivity because it determines whether people can reach and use a location in public spaces. Hidayati et al. (2020) describe intrinsic (individual perceptions and abilities) and extrinsic factors (spatial and sociocultural factors) that influence benefits and burdens for people. Problems are not necessarily caused by individual abilities, but rather by non-inclusive environments created by external factors such as spatial configurations and social norms. This is the difference between potential accessibility and realized accessibility (Hidayati et al. 2020). The mobility of individuals determines their capacity to utilize potential accessibility, resulting in different levels of realization (Hidayati et al. 2021b).

The choice of Niš as a single case study is related to its unique spatial characteristics that incorporate Ottoman influences with socialist city planning in Eastern Europe as part of socialistic industrialization. The city is a mid-sized town in the southern part of Serbia with around 200,000 inhabitants in 2022 (The Statistical Office of the Republic of Serbia 2022a). Originating from early settlements 6,000–5,000 BC, being 400 years under Ottoman rule and after the breakup of former Yugoslavia, Niš is nowadays the third largest city and one of the most important economic centres of Serbia. As a university town, the city attracts a young and dynamic population. However, the Nišava district to which Niš belongs has the third largest amount of households where all members are aged 65 years and older (The Statistical Office of the Republic of Serbia 2022b). Within Niš, this amounts to more than 20,000 households with only older adults according to the recent census. 18% of older adults (65 years and older) experience some form of disability (i.e. seeing, hearing, walking or climbing up the stairs, remembering/concentration, self-care and communication) (The Statistical Office of the Republic of Serbia 2022c). These statistics indicate a greying population that requires accommodation accessing public spaces if quality of life is to be maintained. 30% of children within the city are under the age of 10 and are not yet independently mobile. With 18% of all children being under the age of six, there is a significant amount that will require caregivers to navigate the city (The Statistical Office of the Republic of Serbia 2022d). This means that the city itself should cater to underrepresented groups such as older adults, people with disabilities, young children and parents with strollers or similar equipment.

In recent years, there has been a shift towards prioritizing slower modes of transportation such as walking and cycling, overturning the previous prioritization of private cars in the streets in most European cities. There are several planning concepts that cater to different aspects of inclusivity, such as shared spaces (Appleyard 2021), living streets (Bain et al. 2012) and pedestrian-friendly zones (Lah 2018). However, the city of Niš is still considered car-dominated due to a lack of public infrastructure with most residents reliant on the car for their basic needs (Branković et al. 2016). In this chapter, we will examine the pedestrian accessibility of Niš for inclusivity by comparing potential accessibility through spatial analysis of street networks and realized accessibility as perceived by expert users and citizens of public spaces through semi-structured interviews.

9.2 Method and Data

For this study, a mixed method approach was employed, consisting of (1) spatial analysis of a mathematical street network model through the space syntax method, (2) semi-structured interviews and (3) a photographic survey. The space syntax models provide an understanding of the potential spatial accessibility of streets and route hierarchy based on the centrality of individual street segments. The semi-structured interviews ($n = 5$) provided an in-depth understanding of accessibility challenges and opportunities for pedestrian spaces in the city of Niš. The photographic survey allows us to illustrate urban street profile designs.

9.2.1 Computational Street Network Analysis with Space Syntax

The method of space syntax (Hillier and Hanson 1984; Hillier 1996) allows for the analysis of spatial configurations of street networks and each street's centrality represented by spatial accessibility relative to all other streets in the urban system (Czerkauer-Yamu and Voigt 2011). Space syntax is described as a socio-spatial approach as it is evident that the analytical results correlate well with pedestrian movement (Hillier et al. 1993). We identified through space syntax the normative potential of spatial accessibility for the city of Niš in Serbia.

For this study, we were interested in through movement depicting a route hierarchy when transecting space from origin to destination. In this study, through movement potential is analysed through the measure of Normalized Angular Segment Choice (NACH). We applied radius n and radius 1200 m to contrast car route hierarchy by car ($r = n$) and on pedestrian route hierarchy ($r = 1200$ m) for a 15-min walk.

Route hierarchy allows to represent a cascade from the most used to the least used streets when moving from origin to destination. Streets with high potential through movement or pedestrian route choice (Yamu et al. 2017; Van Nes and Yamu 2021) are assumed with high pedestrian movement (Hillier 2004). For the street network model, we used OpenStreetMap data, accessed in June 2023, more specifically road centre line data. A road centre line map is feasible when computing angular segment choice (Turner 2007).

Angular segment choice is calculated by counting the number of times each street segment falls on the shortest path between all pairs of segments within a selected distance (termed the "radius"). The "shortest path" refers to the path of least angular deviation (namely, the straightest route) through the system (Hillier and Iida 2005). We normalized the raw values for data visualization purposes (for more details, see Hillier et al. 2012; Yamu et al. 2021).

9.2.2 Semi-Structured Interview

Five semi-structured interviews with experts and citizens to compare with the spatial analysis results regarding perceived realized accessibility. The expected differences can provide an understanding of inclusivity to evaluate the accessibility of the city of Niš public spaces and allow for a discussion of the impact of (or lack thereof) inclusive accessibility of public spaces. We interviewed five participants. They include a planning expert from the city of Niš, an urban planner living in Niš, one resident with disability and two visitors. Participants were asked about their own experiences and thoughts on inclusive pedestrian accessibility in public spaces. The interviewees were selected through convenience sampling due to access to participants, their experiences and language barriers. The interviews were transcribed by the authors and anonymized. All interviews were conducted in November and December 2023, and each interview took around thirty to sixty minutes. The interviews were then analysed according to how the participants view inclusivity and accessibility in public spaces, and their experiences of inclusive accessibility in the city with specific positive and negative examples.

9.2.3 Visual Analysis

A visual analysis provides a qualitative assessment of the public realm design and the nature of uses on a micro-scale. The micro-scale is important to reveal to which degree the pedestrian accessibility potential from the normative space syntax analysis is realized (Yamu and Garau 2022). For this study, we were especially interested in the street profiles and their design, and how they impact on accessibility and inclusivity of users. For this study, a photographic survey was conducted to visually analyse street profiles and juxtapose the results to the numerical results of the space syntax analysis. The photographic survey was conducted in June, October, and November 2023.

9.3 Results and Discussion

The results from the mixed method approach reflect a consistent assessment of potential and realized accessibility within the city of Niš from an analysis of the structure of the city that shows through movement potential and route hierarchy and a visual analysis of how accessibility is practised and experienced. This is enriched by the nuanced understanding and perception of inclusive accessibility shared by the residents, experts and users who were interviewed. Next, the results from each method are first introduced and the section will end with a comparison of the results.

9.3.1 Through Movement Potential and Route Hierarchy

The city-wide analysis (Fig. 9.1 left) indicated a route hierarchy mostly for individual motorized vehicular movement, whereas the local analysis (Fig. 9.1 right) indicated the walkability potential for Niš. The city-wide analysis depicts a distinctive main route system following the morphology of the Nišava River and the Nišava Valley terrain. The city is embedded within this valley following a west-to-east direction. This urban growth is reminiscent of cities such as Florence in Italy where topography and the river as a historic trading route are the main determinants of the street network's configuration. The city lacks a strong north–south connection through a main route network. The local analysis depicts a potential 15-min walkable neighbourhoods analysis with high walking potential in the wider core of the city, reaching from the main square *Trg kralja Milana* adjacent to the river to the south of the city along a shopping street, historic streets with cafés, mixed uses and tertiary education facilities. Different functions in the city are accessible via car and on foot. Interestingly, the pedestrian through movement analysis (Fig. 9.1 right) indicates higher potential accessibility to important functions such as administrative (Fig. 9.1, function 6), retail (functions 10 and 11), education (function 7) and healthcare (function 8) locations in the city. However, most of these are more easily accessed by cars with the exception of the pedestrianized shopping streets (functions 4 and 5). The well-connected grid serves potentially high to medium pedestrian movement which is often not fully realized (see Sect. 3.3). The walkable 15-min city model is the pivot for developing walkable cities and neighbourhoods to create inclusive public spaces (Yamu and Garau 2022) with mixed land use. Further, the local analysis shows the modernistic urban development at the fringe of the wider urban core (east and south) and alongside the banks of Nišava River (urban centre to west) with low walkability potential (Fig. 9.1 right).

Fig. 9.1 (left) Normalized angular choice city-wide with $r = n$ and (right) normalized angular choice local analysis with $r = 1200$ m

9.3.2 Understanding and Perceptions of Inclusive Accessibility in Public Spaces in Niš

The semi-structured interviews reveal how inclusive accessibility in Niš is first understood and then perceived by residents, experts and visitors. Participants defined inclusive accessibility in relation to being accepting and non-discriminatory to individuals and groups regardless of socio-demographic characteristics.

> *"Inclusion is the process of including and accepting all individuals, regardless of differences in their cultural, ethnic, socio-economic, gender, religious, sexual, physical, psychological or other characteristics. This means that every individual, regardless of differences, should feel respected and included in society."—(Expert #1)*
>
> *"Inclusiveness for me means above all equality. That public space provides the opportunity for everyone to use it equally."—(Expert #2)*

Inclusive accessibility in public spaces is subsequently related to being able to participate independently or to access public areas with unhindered movement for various groups.

> *"Accessibility refers to the ability to access various institutions and public areas for people who use mobility aids (people with disabilities) but also for people with babies in strollers and old people."—(Resident with a young child)*

The nature of public spaces is also determined by how inclusive it is and is perceived as an expression of basic human rights and reflective of local cultural norms.

> *"I think inclusiveness in public space is important because it is precisely this characteristic that defines a space as public. If a space is not inclusive, it can hardly be characterized as public...All our public spaces reflect the long-standing hospitality mentality of this city."—(Expert #2).*
>
> *"I think it is necessary to ensure unhindered movement for all people in public areas so that there is no discrimination against certain social groups. It is a basic human right."—(Expert #1)*

Although the public spaces of the city are seen as generally inclusive in nature, it is not inclusive for certain user groups.

> *"Inclusivity is not represented in all segments of public space in the city in which I live, because I believe that many public institutions prevent the free movement of people with disabilities or the elderly."—(Resident with a young child)*

Understanding inclusive accessibility in relation to public spaces involves multiple layers of meaning. On one hand, individuals apply moral and ethical considerations to the concept of inclusivity, which guarantees non-discriminatory and unhindered movement for all. Additionally, inclusive accessibility is assessed by the ease with which vulnerable populations can independently access a physical space. It is recognized that while certain spaces may be generally accessible, they are not inherently inclusive, especially for those with disabilities or those who require accommodations beyond conventional standards of access. Given the greying population and the significant number of older adults with disabilities, the issue of inclusive accessibility will be increasingly important in Niš. The participants experience the city as being partially accessible depending on the type of users, the location within the city and weather constraints.

"I would describe it as partially accessible. The main square and the fortress are well-connected to all parts of the city, accessible to everyone, serving as a good example. However, the area next to the river is inaccessible, especially for people with disabilities, and its potential is only partially utilized, particularly in certain points rather than along its entire length."—(Expert #2)

"Most public spaces are accessible for all, but the seasons affect the quality... In icy seasons, many parts of the public space become inaccessible, and in locations outside the main core area of the city, the winter darkness restricts access as well."—(Visitor)

The state of public space infrastructure and the dominance of cars in Niš are suggested as leading to negative experiences of public spaces being accessible.

"As a mother of a newborn, I would point out the lack of sufficient pavement for a child in a stroller to walk freely. Too many parking spaces means a direct reduction in space for babies in strollers to walk around...Another problem is the large number of wrongly parked cars that obstruct the movement of people on the sidewalk."—(Resident with a young child)

"As a negative example, I would mention the streets because many years of bad design and carelessness have created various barriers, not providing equal conditions for the use of all people, especially those with mobility challenges."—(Expert #2)

The lack of accessibility is perceived beyond the street as public space but also into the ingress to and access within public buildings.

"For example, many educational institutions do not have an adequate elevator to transport these people. On the other hand, the streets have adequate space for their unhindered movement."—(Resident with a young child).

"In fact, public spaces are partially accessible...This could not be said for elevators, means of public transport, and especially not for the entrances of cultural institutions."—(Expert #1)

Niš is considered partially accessible, and the participants have a clear understanding of the factors that determine accessibility. Accessibility depends on the abilities and limitations of individual users, the location of public spaces within the city, the predominant mode of transportation on the streets, the condition and maintenance of public space infrastructure, and the weather conditions.

The lack of available sidewalks for those who require extra space, such as young caregivers with strollers, due to private cars parking on them, highlights the potential for improving accessibility. However, this is not simply a matter of implementing sidewalks. It requires a broader transformation in how public spaces and streets are utilized and allocated.

9.3.3 Visual Analysis of Accessibility Experience

Niš provides inviting public squares and streets facilitating pedestrians in certain parts of the city. The main shopping street and pedestrian zones (Fig. 9.2a–c) with mixed land use including cloth stores, supermarkets, cafès, snack bars and restaurants are more than able to cater for high pedestrian volumes during the day and at night. Urban furniture in the main public square *Trg kralja Milana* adjacent to the fortress (see Fig. 9.2a) allows for stopping and resting and water features for cooling during warmer seasons. The public square has a pedestrian-friendly design (Fig. 9.3c). These showcase well-maintained and level surface material with no visual or physical inconsistencies. In addition to promoting social interactions amongst public space users, these public space infrastructures allow for vulnerable groups to take a rest and continue with their mobility journey. In ageing societies, these are invaluable and necessary for older adults and those with mobility aids to increase their accessibility to be able to access more of the city.

A number of streets in Niš have not reached peak inclusivity due to poor pedestrian facility design. This is further influenced by the different climate conditions where no effective tree canopies provide for shading and cooling during hot continental summer months with heat waves. Figure 9.3 shows examples of obstructed sidewalks through parking cars on the sidewalks (Fig. 9.3a) and extensions of buildings and traffic sign poles blocking, occupying and annexing urban space for walking (Fig. 9.3b) and too narrow sidewalks (Fig. 9.3c). The example in Fig. 9.3c depicts a potentially unsafe situation for pedestrians due to a lack of minimum width for walking or waiting for passing traffic safely. This narrow sidewalk connects to a crossing for pedestrians that is faded which does not cater for functional and safe pedestrian movement.

These examples illustrate how pedestrians are pushed to walk in the street space meant for vehicular traffic and encounter challenges in traversing the city streets. Often, sidewalks in Niš are not designed to carry larger volumes of pedestrians and the streetscapes do not function for multiple purposes such as building frontage zones, pedestrian zones, planter/furniture zones and curb zones. Further, Fig. 9.3d illustrates how valuable tree canopies that could serve as visual and auditory buffers, for calming traffic, and cooling are currently not well integrated into the street profile but rather become an obstacle for pedestrians. The negative impacts of the trees on the sidewalk are evident in Fig. 9.3d as the trees are not the only obstacle in the sidewalks with significantly reduced width that reduces the convenience of walking. In addition, the cracks in the sidewalk (due to lack of water for the trees followed by roots spreading resulting in asphalt cracks) produce level changes which can be a safety hazard (tripping) for older adults, mothers with toddlers and pedestrians with mobility impairments. In summary, the usability of sidewalks is affected positively and negatively by surface materials, changes in levels, dimensions of gaps, grates and openings, and visual (in)consistency.

Fig. 9.2 (**a**, top left) Pedestrianized shopping street, (**b**, top right) main public square with urban furniture and (**c**, bottom) water feature

Fig. 9.3 (**a**, top left) and (**b**, top right)—obstructed and blocked sidewalks, (**c**, bottom left) too narrow sidewalks for safe walking, and (**d**, bottom right) trees as (unfortunate) obstacles of sidewalks, a side street parallel to Niš main shopping street

9.4 Triangulating the Results

The city-wide analysis indicated a certain route hierarchy for motorized vehicles and a difference in areas of the city where pedestrian accessibility is optimal such as the public spaces, parks and along the river banks). This is in line with the perception of Expert 2 and the visitor. This is also visible from the visual analysis (see Fig. 9.3a–c) where there is a stark difference between the level and quality of pedestrian spaces and what is present or missing in order to facilitate inclusive accessibility. The level of maintenance and design parameters realization are very high in pedestrianized key public spaces but when the general street space even though they are within the city centre is considered, those levels are not always achieved. The visual analysis shows parts of the city that have grown rapidly and been paved in a haphazard fashion whereby valuable landscape features such as trees are not contributing to the pedestrian experience (see Fig. 9.2a–d).

The analysis of the structure of the city shows potential for walkability and perhaps 15-min neighbourhoods. The functions are spread and potentially accessible on foot but the car continues to dominate the street space. This is shown in how private vehicles either infringe or block pedestrian spaces in the visual analysis (Fig. 9.2a–d). The exception to this is the pedestrianized shopping streets. This is reflected in the information shared by our interview participants who gave multiple examples that showed how while public spaces in Niš are generally seen as inclusive, they are not inclusive for all user groups. The lack of sidewalk availability is a leading concern, particularly for young caregivers with strollers. The lack of continued accessibility is a key factor of influence that accounts for the reliance on car use to navigate and participate in this city.

Another issue lies in the lack of a strong north–south network and the fragmentation of pedestrian spaces along the west–east axis. This is partially a legacy of the terrain and the organic growth of the city similar to other European mediaeval centres. If the fragmentation is addressed and the quality of the linked street spaces is guaranteed, the city will be able to realize its potential accessibility for inclusive pedestrian environments. The experts interviewed suggested that realizing the potential accessibility will require a broader transformation of how public spaces and streets are designed and used. In this, the interview participants are unequivocal in their definition of what would be inclusive accessibility, the access to these main public spaces including those streets that connect to these main squares and parks and even those transition spaces between the building interior and exterior needed to be equally accessible to all groups within the population.

9.5 Concluding Remarks

We set out to understand the potential pedestrian accessibility and realized accessibility of Niš by analysing its structure and street networks. This is further supported by interviews on how users (experts, residents and visitors) experienced the city and evidenced through photographic material that illustrates the spatial character of the streets and public spaces of Niš. This study indicated that the city has a route hierarchy favouring motorized vehicles which has led to compromising pedestrian accessibility with the exception of key public spaces such as pedestrianized zones and parks. There is also a visible and experienced contrast between the quality of pedestrian spaces in public areas and general streets. Most public spaces are seen as inclusive, but not for all user groups, with young caregivers with strollers particularly affected. The city is a typical example of a rapidly growing city that has sacrificed street space for motorized vehicles and therefore negatively impacts pedestrian accessibility suffering from the effects of socialist planning.

The spatial analysis emphasizes the potential for increased walkability and even 15-min neighbourhoods but that would require connecting the somewhat lacking north–south networks and the fragmented pedestrian spaces along the east–west axis. This would require a transformation of public spaces and streets that are more in line with the definitions and understanding of inclusive accessibility as provided by our interview participants.

Future research should look systematically into an evaluation of the quality of the streets and pedestrian experience (i.e. level of maintenance and physical or safety barriers). More insights are also needed into the profiles of vulnerable groups and how they have challenges in experiencing inclusive accessibility within the city. The key takeaway for policymakers is to expand the same high-quality public spaces into a cohesive and connected network of streets. Here, the insights into pedestrian experience will be valuable. Once these spaces are realized, the car dominance can be addressed.

Acknowledgements We would like to express our gratitude to Jelena Stanković for connecting us with urban experts and locals in Niš, whose invaluable insights enriched our study. Our thanks also extend to the assistance of John Östh and Marija Dzunic for valuable feedback. Additionally, we would like to thank all the participants for their valuable time and contribution to this research chapter.

Funding This study was supported by the European Union's Horizon Europe research and innovation programme WIDERA-2021-ACCESS-02 under grant number 101059994.

References

Appleyard B (2021). Livable Streets 2.0. Elsevier

Bain L, Gray B, Rodgers D (2012) Living streets: strategies for crafting public space. John Wiley & Sons

Branković MD, Protić IB, Djekić J, Mitković P (2016) Post-socialist suburbanization and sprawl development patterns-Niš case study. Facta Universitatis, Series, Architecture and Civil Engineering, pp 355–366

Bringa OR (2018) Universal design as a technical norm and juridical term—a factor of development or recession? Stud Health Technol Inform 256:33–39 PMID: 30371458

Clarke J (2004) Access for All? The promise and problems of universalism. Soc Work Soc 2(2):216–224

Czerkauer-Yamu C, Voigt A (2011) Strategic planning and design with space syntax. In: Zupancic T, Juvancic M, Verovsek S, Jutraz A (eds), pp 125–133

Hidayati I, Tan W, Yamu C (2020) How gender differences and perceptions of safety shape urban mobility in Southeast Asia. Transport Res F: Traffic Psychol Behav 73:155–173

Hidayati I, Tan W and Yamu C (2021a) Conceptualizing mobility inequality: mobility and accessibility for the Marginalized. J Plan Lit. 088541222110128

Hidayati I, Yamu C and Tan W (2021b) You have to drive: Impacts of planning policies on urban form and mobility behavior in Kuala Lumpur, Malaysia. J Urban Manag. https://doi.org/10.1016/j.jum.2020.12.004

Hillier B (1996) Space is the machine: a configurational theory of architecture. Cambridge University Press, Cambridge, UK

Hillier B (2004) Can streets be made safe? Urban des Int 9:31–45

Hillier B, Hanson J (1984) The social logic of space. Cambridge University Press

Hillier B, Penn A, Hanson J, Grajewski T, Xu J (1993) Natural movement: or, configuration and attraction in urban pedestrian movement. Environ Plann B Plann des 20(1):29–66

Hillier B, Yang T, Turner A (2012) Normalising least angle choice in Depthmap-and how it opens up new perspectives on the global and local analysis of city space. J Space Syntax 3(2):155–193

Hillier B, Iida S (2005) Network and psychological effects in urban movement. In: International conference on spatial information theory. Springer Berlin, Heidelberg

Lah O (ed) (2018) Sustainable urban mobility pathways: policies, institutions, and coalitions for low carbon transportation in emerging countries. Elsevier

Van Nes, A., Yamu, C., 2021. Analysing linear spatial relationships: the measures of connectivity, integration, and choice. Introduction to space syntax in Urban Studies, pp 35–86

Venter C, Mahendra A, Hidalgo D (2019) From mobility to access for all: expanding urban transportation choices in the global south. World Resour Inst Wash DC 1–48

The Statistical Office of the Republic of Serbia (2022a), Total population, by municipalities and cities. Available at: https://popis2022.stat.gov.rs/media/31359/0_ukupan-broj-stanovnika.xlsx. [Last accessed: 02 Jan 2024]

The Statistical Office of the Republic of Serbia (2022b) Elderly households, by municipality. https://popis2022.stat.gov.rs/media/31371/12_staracka-domacinstva.xlsx. Last accessed 02 Jan 2024

The Statistical Office of the Republic of Serbia (2022c) Population by disability status, age, sex and type of settlement, by municipalities and cities. https://popis2022.stat.gov.rs/media/31415/18_stanovnistvo-prema-statusu-invaliditeta-starosti-polu-i-tipu-naselja.xlsx. Last accessed 02 Jan 2024

The Statistical Office of the Republic of Serbia (2022d) Families with children by family type, number and age of children, by municipalities and cities. https://popis2022.stat.gov.rs/media/31374/15_porodice-s-decom-prema-tipu-porodice-broju-i-starosti-dece.xlsx. Last accessed 02 Jan 2024

Turner A (2007) From axial to road-centre lines: a new representation for space syntax and a new model of route choice for transport network analysis. Environ Plann B Plann des 34(3):539–555

Yamu C, Van Nes A, Garau C (2021) Bill Hillier's legacy: space syntax—a synopsis of basic concepts, measures, and empirical application. Sustainability 13(6):3394

Yamu C, Garau C (2022) The 15-Min City: a configurational approach for understanding the Spatial, economic, and cognitive context of walkability in Vienna. In: International conference on computational science and its applications. Springer International Publishing, Cham, pp 387–404

Yamu C, Poplin A, Devisch O, De Roo G (eds) (2017) The virtual and the real in planning and urban design: perspectives, practices and applications. Routledge

Chapter 10
Urban Bikability: A Case-Comparison of Utrecht, the Netherlands and Nis, Serbia

Daniel Piatkowski, Chaoru Lu, and John Östh

Abstract The built environment plays a significant role in the decision to bicycle for transport purposes. It is also the primary barrier to introducing bicycling as a viable means of transport in developing urban mobility systems. As such, bikability—the relative-friendliness of a transport system to bicycling—is a topic of increasing interest among planners and policy makers. Identifying and quantifying the factors that contribute to a city's bikability is essential to providing an evidence-base for efforts to create multi-modal, sustainable, healthy, and just transport systems. In this chapter, we identify, describe, and compare the built environment characteristics of a high-bicycling city to a low-bicycling city so as to make recommendations for improving the overall quality of the bicycling environment. With the advent of so-called big data (i.e., large amounts of publicly available, spatially referenced, and up-to-date information about the built environment), we have the opportunity to better understand urban environments than was previously possible. The potential is there to use big data and methods to understand urban bikability and provide clear directions for practitioners. This research aims to do just that, situating findings within social and political context, and offering directions for increasing urban bikability internationally.

Keywords Sustainable mobility · Urban cycling · Travel behaviour · Accessibility · Machine learning

10.1 Introduction: The Bicycle and Urban Change

Urban bicycling is experiencing a renaissance in recent years. Beginning with street experiments (Glaser and Krizek 2021) during the pandemic, a bike boom (Piatkowski 2024) has swept across many cities globally as part of larger waves

D. Piatkowski (✉) · C. Lu · J. Östh
Department of Built Environment, Oslo Metropolitan University, Oslo, Norway
e-mail: daniel.piatkowski@oslomet.no

J. J. Stanković et al. (eds.), *Regional and Urban Challenges and Opportunities for Data-Policy Interactions*, The Urban Book Series,
https://doi.org/10.1007/978-3-032-08592-4_10

of interest in human-focused urbanism, like car-free (Nieuwenhuijsen and Khreis 2016) and 15 min cities (Abdelfattah et al. 2022). Taken together, cities are working to become more sustainable, nore just, more livable, and healthier. Doing so means altering approaches to both transport and land use away from auto-centric urbanism and towards human-centric urbanism.

In this context it makes sense that bicycles are booming. Bicycles are particularly well-suited to the role of transitional transport—a bridge between auto-centric cities and human-centric cities. Bicycles can easily substitute for short trips by cars (Piatkowski et al. 2015). Modern cargo bicycles can carry families (Thomas 2022) and increasingly handle urban freight and delivery (Sheth et al. 2019). And e-assist technology has become so successful that e-bikes are beginning to outsell electric cars in some places (Zipper 2024). Taken together, the bicycle (in all its modern forms) is a viable car-replacements in more circumstances than ever before and thus has the potential to transform cities more ways than previously thought possible.

But a human-centered city needs to be about more than mode shift. Mode shift is the first step; basic bike infrastructure is the first step. From there, the surrounding urban environment can change to become more vibrant and accessible. Accessibility—the ability to access daily needs—needs to be increased for non-auto modes to be viable (Handy 2020). Bicycles, in addition to their potential to substitute for driving, also have the ability to jumpstart street-level accessibility increases. These increases in turn can be contagious, leading to neighborhood and city-scale changes in accessibility.

The process is a simple one. Investing in basic bicycle infrastructure can induce more cycling (Dill and Carr 2003). When cycling, it is easy to stop at local destinations throughout daily and weekly travel, and research demonstrates that people on bikes stop more often and spend more money per trip than those in cars (Popovich and Handy 2014). Once businesses begin to cater to cyclists, they spur more bicycle traffic. Bicycle traffic in turn slows cars and make streets safer for everyone (Nordback and Marshall 2012). Over time, bicycle corridors can transition into accessible, multimodal corridors. In this way, investing in bicycling spurs large-scale changes in cities.

The power of bicycles to spur urban change is well-documented in high-bicycling cities. Global cycling capitols like Amsterdam in the Netherlands or Copenhagen in Denmark are also well-known for being very sustainable, healthy, and livable places (Koglin et al. 2021). This is no accident. Investing in bicycling supports these larger efforts at urban change. This ability of bicycles to complement larger efforts at urban sustainability explains the bicycle renaissance in cities today. It also explains why bicycle-friendliness has become an urban planning goal in its own right (Koglin and Rye 2014).

In this chapter, we explore bikability—the relative-friendliness of a transport system to bicycling—as a key metric to urban sustainability. We do so using a combination of novel methods and so-called big data sources. We map the street networks of our comparison cities; we then use Google Streetview images along the street networks to create a database of street-level imagery. Finally, we apply machine learning to quantify street-level built environment factors and statistical analyses to assess bikability. We do this in two very different cities to present a case comparison of bikability in a high and low-bicycling city. Comparing findings across these two cities allows us to better understand the implicit and explicit approaches to transport and land use planning that resulted in these distinct outcomes. Finally, we discuss the implications of these findings and, in particular, of these two distinct approaches to urban mobility to build an evidence-base for better urban futures.

10.2 Background: Bikable City Planning

Making a city more bikable is a process involving institutional, physical, social, and cultural evolution. It starts with redefining the culture around transportation and the goals of transportation—from a speed-centric approach to one that considers holistic impacts and potential for transport in cities (Tranter and Tolley 2020). In cities that have traditionally been dominated by the automobile, that process is particularly difficult (Koglin and Rye 2014). This is because bicycle-centric transportation systems favor, and foster, livability through accessibility. In contrast, auto-centric transportation prizes mobility—the rapid movement between destinations—over access to destinations (Handy 2005). This is why bicycle-centric cities are denser and more vibrant, while auto-centric cities are sprawling and devoid of street life.

The literature on bikable places overlaps with multiple aspects of planning and policy: the built environment, the macro- and micro-level environment, travel behavior, and urban accessibility. In general, this literature is correlational but points the volume of correlational research underscores the robustness of certain trends and best practices for cities seeking to become more bikable.

There is a clear relationship between travel and the built environment (Ewing and Cervero 2010). The study of the built environment's relationship to travel behavior includes numerous sub-fields in land use and transportation studies. Ewing and Cervero, in two meta-analyses, confirmed the significance of factors including density, diversity, design, destination accessibility, and distance to transit all impact mode choice and travel behavior (Ewing and Cervero 2010; 2001). Finding that denser, more diverse built environments are associated with non-auto travel. At the macro-scale, accessibility studies—the study of location-based access to daily needs—confirms that high accessibility places are associated with walking, bicycling, and transit use. In turn, low accessibility places are associated with auto-reliance.

Travel behavior research further confirms that both micro- and macro-built environment factors significantly impact travel behavior and mode choice decisions (Mertens et al. 2015). Macro-level-built environment characteristics include traffic and congestion, network connectivity and accessibility, safety (perceived and actual), and policies targeted at commute mode choice. At the micro-level, myriad factors impact travel behavior, including street-level characteristics such as street design and surrounding buildings, traffic speed and volume, multi-modal infrastructure type and quality, and vegetation (Müggenburg et al. 2022).

Micro- and macro-scale findings regarding the relationship between the built environment and travel behavior demonstrate that as the density of destinations (i.e., accessibility) increases, and the quality of street design and infrastructure increases, so does bicycling (Mertens et al. 2015; Müggenburg et al. 2022). Taken together, this confluence of forces creates places that are more inviting on a human-scale, and in turn lead to multiple additional benefits for people and the planet.

10.3 Significance: Why Bicycles Matter

Transportation is a leverage point that can be utilized to make cities more sustainable and resilient, more equitable and just, and more healthy and livable (Deakin 2007). Transportation emissions account for approximately 30% of emissions produced by cities. That percentage does not account for the carbon footprints of sprawling and auto-oriented development (Sarkodie et al. 2020). Bicycles are a clean and sustainable form of transportation, and bicycle-friendly built environments reduce the carbon footprint of cities (Chen et al. 2022).

Bicycles are also an important tool in addressing transportation equity. Private vehicles are inherently inequitable because of their high up-front and operating costs (Lutz 2014). Auto-centric development is also inequitable because it reduces the feasibility of more equitable modes of transportation and forces society to subsidize vehicle space on streets and parking (Shoup 2021). Furthermore, vehicles and auto-oriented cities are more dangerous from a traffic safety perspective than bicycle-friendly cities (Pucher and Dijkstra 2000). In contrast, bicycles are more efficient and inexpensive than vehicles. Bicycle-centered development is more economically efficient, and bicycles (along with bicycle infrastructure) are associated with reduced road crashes and fatalities (Bopp et al. 2018).

Bicycles, and bicycle-friendly cities, also have the potential to improve public health. Bicycles are an active mode of transportation that can increase daily physical activity (Donaire-Gonzalez et al. 2015). Unlike vehicle traffic, bicycles do not negatively impact urban air quality (Johansson et al. 2017). In contrast, cars (and increasingly heavier, electric vehicles) are impacting particulate matter levels in cities through wear on tires and roads (Timmers and Achten 2016). Bicycles, and bicycle-friendly cities, are an antidote to urban air quality concerns.

10.4 Context: Alternative Approaches to Transport

This chapter presents a comparative case study of two very different cities, focusing on bicycle-friendliness. The bicycle can be a change agent for cities, but it is also an indicator species: by exploring urban bikability, we can better understand the context, approach and goals of transport and land use planning that resulted in very different degrees of bicycle-friendliness. The Dutch are of course well-known for bicycle-centric planning, but in contrast, Serbia and the Western Balkans have planned their transport systems around the automobile. These two approaches are themselves the result of very different histories, particularly in recent years. By examining both in parallel, we make recommendations for more sustainable urbanism in the future.

The Dutch are global leaders in urban bikability. Beginning in the 1970s, the Dutch began transitioning their transport and land use systems to be more sustainable and multi-modal. This began with grassroots advocacy efforts by a coalition of community members, parents of young children, safe streets advocates, cyclists, and preservationists. This broad coalition was explicitly against the rapid change in the post WWII years of Dutch cities from cities designed around people, to cities designed to accommodate a growing fleet of single occupancy vehicles (Dekker 2022). These early advocacy efforts created momentum around change, but on their own were not successful at stemming the tide of auto-dominant design and planning.

It took a crisis beyond their borders to push the Dutch down the path of sustainable urbanism and world-class bikability: the 1970s oil crisis. Almost immediately, the Dutch government realized auto-oriented cities meant cities reliant on foreign oil and pivoted accordingly. Suddenly, the goals of government priorities and grassroots advocates aligned. Advocates wanted cities for people, traversible by human-powered transport. The government wanted inexpensive transport infrastructure that did not depend on fossil fuels. As such, bikability became the priority (Reid 2017).

In the preceding decades, Dutch cities have become the gold-standard for human-centered urbanism. Over that time, Dutch design, planning, and policy has shifted to make bikeable cities the standard. This was revolutionary in the 1970s through the 1990s, when auto-oriented urbanism became entrenched in much of the developed world (particularly in North America and Australia, but also throughout much of Europe), and was also exported to developing countries and the Global South in the form of international aid and infrastructure programs. What that means is that today, turning back the clock on auto-centric urbanism is a priority for cities globally. And these cities often look to the Dutch for guidance.

Dutch guidance on bikable urbanism is plentiful. As described above, accessibility and both macro- and micro-built environment factors matter for bikability. In addition to these, the Dutch have spearheaded an approach to street design that allows for consideration of all users of the transport system and can be applied broadly across settings, from high-traffic arterials to low-traffic neighborhoods. Speed and traffic volume become the key indicators guiding the type and extent of bicycle infrastructure needed. High speed/high car-traffic streets require the most extensive infrastructure, and as speed and vehicle traffic decreases, less infrastructure is needed. At one extreme are bicycle facilities completely separate from vehicles. At the other extreme are shared streets, very low-speed streets that require no infrastructure to differentiate space for pedestrians, cyclists, or drivers (Bopp et al. 2018). This approach ensures urban bikability while guiding infrastructure investments.

Serbian transport policy is at a crucial juncture, a turning-point as cities and town within Serbia work to develop intelligent, efficient, and equitable transportation systems in the context of rapid regional development and the ever-present challenge of mitigating climate change impacts of transport systems. Cities of former European socialist countries have struggled since 1989 (Mrkajić and Anguelovski 2016). Policies aimed at encouraging foreign investment and economic growth have, in turn, weakened public authority in the planning process. It has also led to short-term thinking in urban planning wherein sustainability, livability, and health are deprioritized in favor of market gains. Today, cities like Nis are grappling with the twin challenges of decades of (i) unsustainable development and (ii) a political-and-technical system that has evolved to facilitate it.

But history is not destiny. Despite unsustainable development practices and a focus on automobile oriented transport systems (Nedović-Budić and Cavrić 2006), increasing internationalization in Serbian planning thought and practice places Serbian cities at an inflection point. Can Serbian urbanism become more sustainable and human-centric? The literature indicates that a focus on accessibility and multi-modal transport systems can set cities like Nis on a path to become regional leaders in bicycle urbanism. To best leverage existing resources, knowledge, and capacity, transport policy must be targeted to leverage points in existing institutional systems and physical spaces.

By comparing Nis, a low-bicycling city, to Utrecht, a high-bicycling city, we can begin to elucidate the existing differences and potential leverage points to enact change. We first describe existing physical characteristics of cities and in turn link them to planning cultures and histories that have evolved in concert with the environments they govern (Stead et al. 2015). We can then make evidence-based recommendations for defeating institutional path dependence (Sorensen 2015) and setting a course for a brighter urban future in Nis.

10.5 Comparative Case Study: Data and Methods

We develop comprehensive maps of the built environment for two case study cities, one with high levels of bicycling (Utrecht) and one with low-bicycling levels (Nis). Rather than a traditional mapping exercise resulting in an analysis at the "birds-eye-view," this work quantifies bikability from the street level, or "bicyclist's-eye-view." To do this, we have created 360 views at the street level of the road networks in these two cities using Google Streetview images. We then apply machine learning to identify and quantify the properties of the built environment. The resulting dataset forms the basis of our comparative research.

We use GIS material from Open Street Map (OSM) and Google Streetview images for this study. The OSM was used to identify and distinguish cycle roads from the road network in both cities. In Figs. 10.1 and 10.2, the data and methods used to generate variables are illustrated. Each figure has four parts in which part A illustrates locations where used Google Streetview images are located for which 360° imagery has been captured and analyzed using ML. Part B illustrates the road network in the area (including cycle infrastructure). Parts C and D shows output from GIS analyses where the density of roads within a radius of 500 m from all locations

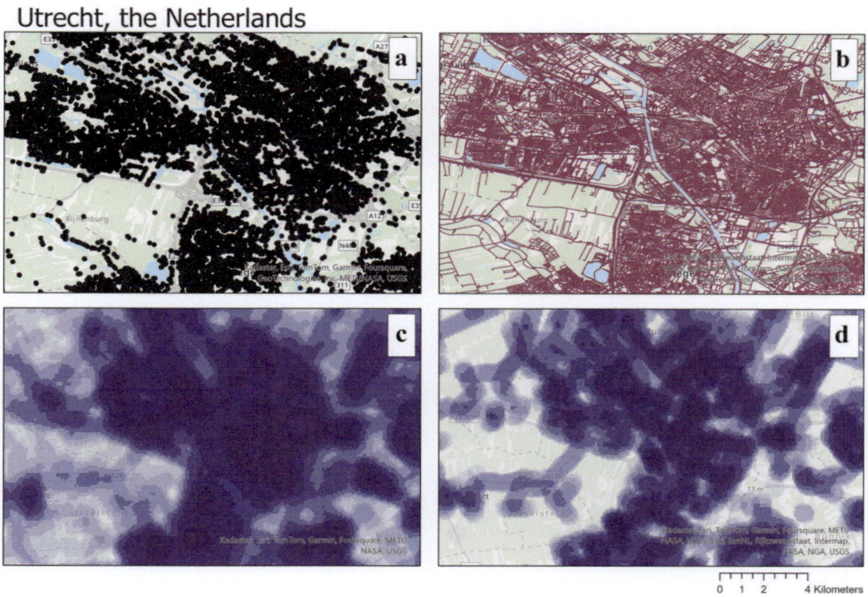

Fig. 10.1 Depicting the locations of Google Streetview images used in the ML analyses in Utrecht (part A). In part B, the street network is illustrated, and in parts C and D, the density of roads (C) and cycle roads (D) are illustrated

(determined as 10 m grid cell units) are shown, and where C indicates the density of all roads, while D indicates the density of cycle infrastructure.

10.6 Case Study Findings

In this section we present and discuss how the density and distribution of cycle infrastructure is associated with features identified in the 360° images from the two studied cities, using regression analysis and comparing means. Prior to the regressions, we provide tables detailing the distribution of features in the two cities and how they relate. In Table 10.1, details from the ML analyses of the images are revealed. In the count columns, the number of captured features in the two cities is shown. In both cities, roads and sidewalks are unsurprisingly the most common features captured, while features such as rider (i.e., sitting on a motorcycle or on a bike) are less frequent. The mean and std columns are capturing averages and standard deviations of the ratios generated for all images, which means that a greater average ratio suggests that the observed feature is covering an on average corresponding fraction of the feature in the images.

The descriptives in Table 10.1 suggest that there are only minor differences between the cities; however, if we employ an independent sample T-test to compare

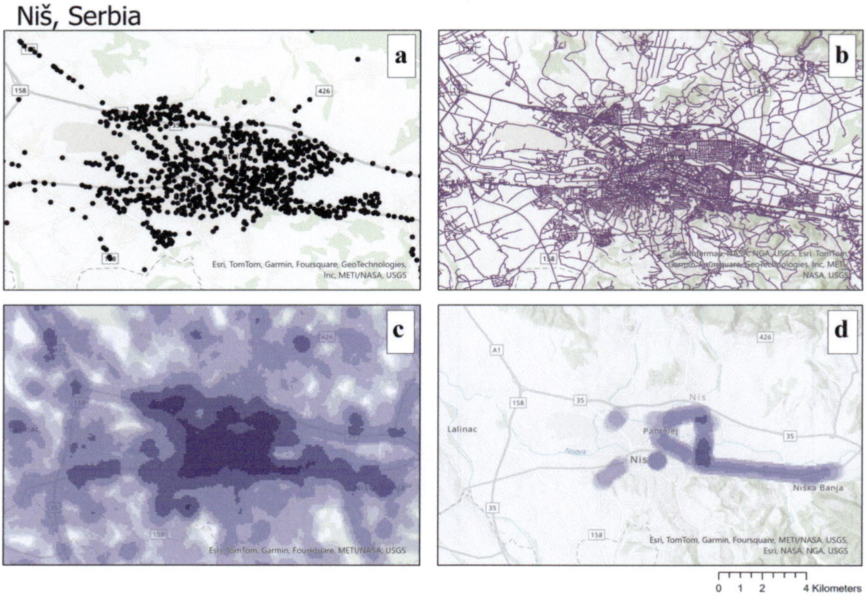

Fig. 10.2 Depicting the locations of Google Streetview images used in the ML analyses in Nis (part A). In part B, the street network is illustrated, and in parts C and D, the density of roads (C) and cycle roads (D) are illustrated

the occurrence of features in the two cities, we find significant differences (see Table 10.2). With positive values indicating greater presence in Utrecht (and vice versa for Nis), significant differences are found for most features and strong and significant differences are found for sidewalks, poles, and bicycles being more common in Utrecht, and terrain, sky, and walls being more common in the Nis images. The statistics so far is just telling us about the distribution of features in the urban landscape. In our next analytical step, regressions will indicate if these features are more or less frequent in areas where bicycle infrastructure is located.

10.7 Research Results

Using the share of cycle roads among all roads within a radius of 500 m from all locations where there is machine learning imagery available in the two urban areas, we have generated a dependent variable that can be used to analyze the surroundings of areas where cycling, or at least infrastructure for cycling, is more or less frequent.

In Table 10.3, coefficients of the machine learning variables are shown for both cities separately. For both Utrecht and Nis, roads are significantly more visible in the cycle road-dense areas. This makes sense if we consider that density of roads and

Table 10.1 Machine learning analysis

	Utrecht			Nis		
	Count	Mean	Std	Count	Mean	Std
Road	17,682	0.194	0.087	2535	0.187	0.092
Sidewalk	17,665	0.067	0.051	2538	0.042	0.038
Building	17,691	0.146	0.138	2535	0.141	0.158
Wall	17,523	0.012	0.018	2506	0.020	0.034
Fence	17,571	0.020	0.031	2418	0.016	0.021
Pole	17,421	0.004	0.003	2457	0.003	0.004
Traffic_light	9666	0.000	0.001	1451	0.000	0.001
Traffic_sign	14,676	0.002	0.003	1307	0.001	0.003
Vegetation	17,673	0.278	0.172	2538	0.260	0.171
Terrain	16,947	0.022	0.031	2504	0.042	0.040
Sky	17,523	0.240	0.099	2528	0.306	0.112
Person	7784	0.001	0.005	810	0.001	0.003
Rider	2650	0.001	0.002	266	0.001	0.002
Car	16,172	0.018	0.025	2256	0.016	0.028
Truck	6878	0.003	0.014	1079	0.004	0.028
Bus	5589	0.002	0.012	1059	0.001	0.006
Train	9959	0.001	0.010	1933	0.002	0.008
Motorcycle	10,250	0.001	0.002	1132	0.001	0.002
Bicycle	15,109	0.001	0.004	2048	0.001	0.002

crossings could be perceived as indicators of dense traffic areas where specific solutions for non-motorized vehicles could be needed. In Utrecht, pedestrian dominated sidewalks become more common near cycle roads, further suggesting that traffic separate lanes on the basis of mode. Similarly, buildings, walls, fences, and poles are more common near cycling areas in Utrecht. In Nis, the patterns are different. With the exception for poles also being positively associated with cycling, the positive and significant coefficients shown for Utrecht are either insignificant or in the case of walls, negative.

These results are likely consequences of the differences in localization of cycle infrastructure in the two urban areas. As also shown in Figs. 10.1 and 10.2, cycle infrastructure is commonplace in all parts of Utrecht, while in Nis it is concentrated in a few areas outside the immediate city center. This also explains why trucks are more strongly and positively associated with cycle infrastructure in Nis and why traffic lights are negatively correlated (traffic lights are more common in denser parts of the cities).

By focusing on the natural surroundings in the two urban areas, we see that terrain and sky are positively and significantly associated with bike-lanes in Utrecht, but insignificant in Nis. Vegetation is also insignificant in Nis (and excluded due

Table 10.2 Independent samples T-test

Variable	T_stat	p_value
Road	3.743	0.000
Sidewalk	23.763	0.000
Building	1.759	0.079
Wall	−18.747	0.000
Fence	5.646	0.000
Pole	12.403	0.000
Traffic light	2.723	0.006
Traffic sign	8.181	0.000
Vegetation	5.039	0.000
Terrain	−28.512	0.000
Sky	−31.058	0.000
Person	−0.836	0.403
Rider	0.336	0.737
Car	3.230	0.001
Truck	−2.080	0.038
Bus	1.077	0.282
Train	−1.062	0.288
Motorcycle	1.133	0.257
Bicycle	9.310	0.000

to model collinearity in Utrecht). What does this mean, and what do the variables depict from a cycling perspective? In separate analyses (not included in the chapter), we find that sky and terrain are positively and significantly correlated with each other, and that vegetation is significantly and negatively associated with sky, terrain as well as with cycle-lanes in both cities. Terrain and sky are more frequent in areas where farming is more common, and vegetation is more common in locations with forestry. The regression results therefore suggest that cycle infrastructure is unusual in forested areas, and that cycle infrastructure in proximity to farmland is common in Utrecht but insignificant in Nis.

The remainder of the variables mostly renders insignificant results, which makes sense since the variables capture the presence of vehicles and people on the street that likely are similarly distributed throughout the city landscape, with one important exception: bicycles should ideally be significantly more common in the cycle infrastructure dense parts of the cities, and while there is a positive correlation, the correlation is insignificant. Interestingly, the strongest, yet insignificant effect, is found in Nis. A possible explanation might be that since cycle infrastructure is common throughout Utrecht the omnipresence of bikes dominates the city image regardless if we have designated lanes or not; and in contrast, the scarcity of bikes in Nis may partially be related to the lack of designated cycle infrastructure, and the stronger coefficients my indicate that the few bikes that roam around in Nis are more

Table 10.3 Regression results

	Utrecht	Nis
Road	0.069 (0.006)***	0.063 (0.017)***
Sidewalk	0.114 (0.009)***	−0.021 (0.023)
Building	0.019 (0.004)***	0.026 (0.016)
Wall	0.055 (0.024)*	−0.041 (0.02)*
Fence	0.076 (0.013)***	0.059 (0.032)
Pole	0.512 (0.131)***	0.513 (0.154)***
Traffic_light	1.22 (0.685)	−2.362 (1.05)*
Traffic_sign	−0.168 (0.16)	0.47 (0.234)*
Terrain	0.186 (0.015)***	−0.014 (0.022)
Sky	0.023 (0.005)***	0.005 (0.018)
Person	0.105 (0.134)	0.45 (0.352)
Rider	0.457 (0.453)	−1.268 (0.804)
Car	0.009 (0.018)	0.017 (0.026)
Truck	0.103 (0.05)*	0.203 (0.031)***
Bus	0.096 (0.067)	−0.146 (0.145)
Train	0.065 (0.06)	0.081 (0.084)
Motorcycle	−0.35 (0.226)	0.786 (0.398)*
Bicycle	0.047 (0.127)	0.566 (0.315)
Vegetation	Excluded due to collinearity	0.011 (0.016)
Constant	0.087 (0.002)***	−0.005 (0.017)
Adj R^2	0.034	0.066
F	35.764	10.417

* = p <.05
** = p >.01
*** = p >.001

frequent along cycle roads but not in any number that would affect the significance of the model.

10.8 Discussion

This research explores urban bikability across two very different case study cities. The empirical results yield generally expected findings, but also highlight opportunities moving forward. Results support the fact that Utrecht has a robust, comprehensive, and mature multi-modal transport system; that is, Utrecht is a world-class example of a highly bikable city. Utrecht today is the result of decades of planning and policy efforts to reduce auto-reliance and create a highly accessible, sustainable, equitable, and livable place.

In contrast to Utrecht, Nis is a city in transition, in a nation (and region) in transition. Cities across Serbia and the Western Balkans are at an impasse, with multiple options to consider and in need of an evidence-base to inform those decisions. Cities in the region are grappling with the question of how to realize sustainability and livability goals in ways that are both cost effective and fast. The empirical analysis suggests clear steps forward.

Before making recommendations regarding transport and land use planning, it is important to recognize certain limitations of this work. One of the goals of this work is to demonstrate the application of big data and new methods to urban bikeability. This research used widely available spatial data about the urban environments in Nis and Utrecht, and when paired with streetview imagery and machine learning, offers clear descriptives of each case's urban environment. Furthermore, this work is novel in that it assesses bikability from the street level, that is, the "bicyclist's-eye-view," as opposed to a "bird's-eye-view" more common in spatial analysis. Further research using this approach could be used to yield specific, street-level urban design, traffic, and infrastructure guidance. Additionally, this approach can be applied broadly and with very little cost to understand the characteristics of cities, streets, and their implications for transport.

However, using big data can be problematic. The usual concerns associated with secondary data sources are amplified when using big data sources. There are systemic biases in data collection and processing that occur before researchers gain access to the data and can be difficult to detect with big data. At its worst, such limitations of big data can lead to flawed findings that exacerbate existing inequities or unjust outcomes in cities. Furthermore, researchers must be cautious when using big data in statistical analyses as large sample sizes can often make it difficult to distinguish between significance and effect sizes. Taken together, big data can be valuable to urban analytics, but is not a panacea. Robust research requires theoretical foundations, robust design, and replication.

10.9 Conclusion

Despite certain limitations, this work provides useful takeaways regarding planning policy and practice to reduce auto-reliance and foster urban sustainability. The first consideration is the distinction between land use and transport infrastructure in our case study cities. Both cities are historic, predating the automobile by centuries, and have inherited a certain level of basic accessibility. This is important because it reduces the barriers to human-centric urbanism when each city's history is that of a human-centric place.

Mixed land uses that encourage accessibility are essential to bikability, and both cities possess some degree of land use mix. This in stark contrast to younger cities in places like North America or Australia, characterized by single-use land development policies and urban sprawl. This foundation of accessibility, or at least the potential for increasing accessibility without dramatically altering land use, means urban change can occur much faster than in more auto-dominant places.

The similarities between the two case studies cities end with the start of the automobile era. Utrecht and Nis have followed very different approaches to transport planning (and associated infrastructure development) in the modern era. The Dutch approach to transport planning since the 1970s has been explicitly oriented at moving from auto-centric to human-centric urbanism. Bicycles are the preferred mode because they offer a clear substitute for driving and bicycle infrastructure is much cheaper than public transport.

At the street level, the Dutch approach means giving preference to non-auto modes, and even on high-vehicle-traffic streets, treating bicycles as equal to other modes. Recall the Dutch design street philosophy in which higher speed differentials necessitate more extensive infrastructure, while lower speed differentials require less infrastructure. There is clear guidance across all types of streets (e.g., arterial or neighborhood street) and institutional support to fund multi-modal infrastructure. Utrecht today features cycle infrastructure of various types on nearly all streets. This is a direct outgrowth of a transport planning philosophy and street design standard that has evolved over the last 50 years.

Nis, in contrast, exhibits a legacy of auto-centric urbanism. Despite land uses that could easily support multi-modal accessibility, the city is hampered by street design (and an underlying transport philosophy) that does not treat all modes equally. It is clear from the bicycle infrastructure map that Nis has very limited bicycle infrastructure, and that is primarily along a single arterial roadway. This suggests that cycle infrastructure has been placed in only the most dangerous corridor of the city. Cyclists need safe infrastructure that shelters them from arterial traffic, and Nis has at least one example of this. However, this approach also reflects the extent to which cycling is treated as an outlier in the transport system. If cycling was considered a viable mode of transport in Nis, then more extensive infrastructure would be available. In its current form, the bicycle infrastructure in Nis provides minimal protection for those on bicycles, and is not comprehensive, meaning very few people are willing to ride.

But Nis, and many cities in the region, hold a great deal of potential for rapid change. The greatest strength exhibited is a baseline level of urban density and accessibility. Nis can make a great deal of progress by focusing on street-level urban change and simple traffic policies. By following the Dutch example, Nis can become safer and more multi-modal by reducing vehicle speed limits and increasing the cost of parking, while simultaneously increasing the frequency of public transit and build-out basic bicycle and pedestrian infrastructure that connects destinations to the existing (limited) bike system. The existing bicycle path can serve as a tree trunk by which additional infrastructure can branch off. Infrastructure can be paired with policies to reduce traffic speeds to make streets safer. As streets become safer and infrastructure more extensive, more people may feel comfortable trying to ride.

Sustainable and livable urbanism can begin with modest investments in bicycling. These investments can begin a change process that works across culture, policy, and the built environment. Culture change happens by building a culture of cycling. That means making cycling a mainstream activity for people of all ages and abilities. This in turn spurs a policy environment that supports the needs of a growing number of travelers by non-auto modes. In turn, the built environment can continue to evolve to become more human-centered.

The path to urban bikability and therefore more livable and humane places is clear. Utrecht is a valuable case study for cities to follow. Nis has significant potential to become more bikable and in so doing become a regional center for vibrant urbanism, with accompanying social and economic benefits. But Nis is also a good case study for many cities globally that have not yet decided on a course forward. There are political barriers, economic concerns, and competing priorities. While this work presents exciting opportunities to use novel data and methods to inform urban bikability, no amount of analytics can defeat political deadlock. Making bikability a key priority means convincing the public and policymakers of its benefits, of which there are many. The barriers to achieving these ends are political and not technical.

Funding This work was supported by the European Union's Horizon Europe research and innovation program WIDERA-2021-ACCESS-02 under grant number 101059994.

References

Abdelfattah L, Deponte D, Fossa G (2022) The 15-minute city: interpreting the model to bring out urban resiliencies. Transp Res Procedia 60:330–337. https://doi.org/10.1016/j.trpro.2021.12.043

Bopp M, Sims D, Piatkowski D (2018) Bicycling for transportation: an evidence-base for communities. Elsevier

Chen W, Carstensen TA, Wang R, Derrible S, Rueda DR, Nieuwenhuijsen MJ, Liu G (2022) Historical patterns and sustainability implications of worldwide bicycle ownership and use. Commun Earth Environ 3(1):1–9. https://doi.org/10.1038/s43247-022-00497-4

Deakin E (2007) Equity and environmental justice in sustainable transportation: toward a research agenda. https://escholarship.org/uc/item/9607n3t9

Dekker H-J (2022) Cycling pathways: the politics and governance of Dutch cycling infrastructure, 1920–2020. Amsterdam University Press. https://doi.org/10.5117/9789463728478

Dill J, Carr T (2003) Bicycle commuting and facilities in major U.S. cities: if you build them, commuters will use them. Transp Res Rec 1828(1):116–123. https://doi.org/10.3141/1828-14

Donaire-Gonzalez D, de Nazelle A, Cole-Hunter T, Curto A, Rodriguez DA, Mendez MA, Garcia-Aymerich J, Basagaña X, Ambros A, Jerrett M, Nieuwenhuijsen MJ (2015) The added benefit of bicycle commuting on the regular amount of physical activity performed. Am J Prev Med 49(6):842–849. https://doi.org/10.1016/j.amepre.2015.03.036

Ewing R, Cervero R (2001) Travel and the built environment: a synthesis. https://doi.org/10.3141/1780-10

Ewing R, Cervero R (2010) Travel and the built environment: a meta-analysis. J Am Plann Assoc 76(3):265–294. https://doi.org/10.1080/01944361003766766

Glaser M, Krizek KJ (2021) Can street-focused emergency response measures trigger a transition to new transport systems? Exploring evidence and lessons from 55 US cities. Transp Policy 103:146–155. https://doi.org/10.1016/j.tranpol.2021.01.015

Handy S (2005) Planning for accessibility: in theory and in practice. In: Levinson DM, Krizek KJ (eds) Access to destinations. Emerald Group Publishing Limited, pp 131–147. https://doi.org/10.1108/9780080460550-007

Handy S (2020) Is accessibility an idea whose time has finally come? Transp Res Part d: Transp Environ 83:102319. https://doi.org/10.1016/j.trd.2020.102319

Johansson C, Lövenheim B, Schantz P, Wahlgren L, Almström P, Markstedt A, Strömgren M, Forsberg B, Sommar JN (2017) Impacts on air pollution and health by changing commuting from car to bicycle. Sci Total Environ 584–585:55–63. https://doi.org/10.1016/j.scitotenv.2017.01.145

Koglin T, Rye T (2014) The marginalisation of bicycling in Modernist urban transport planning. J Transp Health 1(4):214–222. https://doi.org/10.1016/j.jth.2014.09.006

Koglin T, Te Brömmelstroet M, Van Wee B (2021) Cycling in Copenhagen and Amsterdam. In: Buehler R, Pucher J (eds) Cycling for sustainable cities. The MIT Press, pp 347–370. https://doi.org/10.7551/mitpress/11963.003.0022

Lutz C (2014) Cars and transport: the car-made city. In: A companion to urban anthropology. John Wiley & Sons, Ltd., pp 142–153. https://doi.org/10.1002/9781118378625.ch8

Mertens L, Cauwenberg JV, Ghekiere A, Holle VV, Bourdeaudhuij ID, Deforche B, Nasar J, de Weghe NV, Dyck DV (2015) Does the effect of micro-environmental factors on a street's appeal for adults' bicycle transport vary across different macro-environments? An experimental study. PLoS ONE 10(8):e0136715. https://doi.org/10.1371/journal.pone.0136715

Mrkajić V, Anguelovski I (2016) Planning for sustainable mobility in transition cities: cycling losses and hopes of revival in Novi Sad, Serbia. Cities 52:66–78. https://doi.org/10.1016/j.cities.2015.11.029

Müggenburg H, Blitz A, Lanzendorf M (2022) What is a good design for a cycle street?—User perceptions of safety and attractiveness of different street layouts. Case Stud Transp Policy 10(2):1375–1387. https://doi.org/10.1016/j.cstp.2022.04.021

Nedović-Budić Z, Cavrić B (2006) Waves of planning: a framework for studying the evolution of planning systems and empirical insights from Serbia and Montenegro. Plan Perspect 21(4):393–425. https://doi.org/10.1080/02665430600892146

Nieuwenhuijsen MJ, Khreis H (2016) Car free cities: pathway to healthy urban living. Environ Int 94:251–262. https://doi.org/10.1016/j.envint.2016.05.032

Nordback KL, Marshall WE (2012) Improving bicycle safety with more bikers: an intersection-level study, pp 135–146. https://doi.org/10.1061/41148(389)12

Piatkowski D (2024) Bicycle city: riding the bike boom to a brighter future. Island Press

Piatkowski DP, Krizek KJ, Handy SL (2015) Accounting for the short term substitution effects of walking and cycling in sustainable transportation. Travel Behav Soc 2(1):32–41. https://doi.org/10.1016/j.tbs.2014.07.004

Popovich N, Handy SL (2014) Bicyclists as consumers: mode choice and spending behavior in Downtown Davis, California. Transp Res Rec 2468(1):47–54. https://doi.org/10.3141/2468-06

Pucher J, Dijkstra L (2000) Making walking and cycling safer: lessons from Europe. Transp Q 54:25–50

Reid C (2017) How the Dutch really got their cycleways. In: Reid C (ed) Bike boom: the unexpected resurgence of cycling. Island Press/Center for Resource Economics, pp 179–210. https://doi.org/10.5822/978-1-61091-817-6_9

Sarkodie SA, Owusu PA, Leirvik T (2020) Global effect of urban sprawl, industrialization, trade and economic development on carbon dioxide emissions. Environ Res Lett 15(3):034049. https://doi.org/10.1088/1748-9326/ab7640

Sheth M, Butrina P, Goodchild A, McCormack E (2019) Measuring delivery route cost trade-offs between electric-assist cargo bicycles and delivery trucks in dense urban areas. Eur Transp Res Rev 11(1):11. https://doi.org/10.1186/s12544-019-0349-5

Shoup D (2021) High cost of free parking. Routledge. https://doi.org/10.4324/9781351179539

Sorensen A (2015) Taking path dependence seriously: an historical institutionalist research agenda in planning history. Plan Perspect 30(1):17–38. https://doi.org/10.1080/02665433.2013.874299

Stead D, de Vries J, Tasan-Kok T (2015) Planning cultures and histories: influences on the evolution of planning systems and spatial development patterns. Eur Plan Stud 23(11):2127–2132. https://doi.org/10.1080/09654313.2015.1016402

Thomas A (2022) Electric bicycles and cargo bikes—tools for parents to keep on biking in auto-centric communities? Findings from a US metropolitan area. Int J Sustain Transp 16(7):637–646. https://doi.org/10.1080/15568318.2021.1914787

Timmers VRJH, Achten PAJ (2016) Non-exhaust PM emissions from electric vehicles. Atmos Environ 134:10–17. https://doi.org/10.1016/j.atmosenv.2016.03.017

Tranter P, Tolley R (2020) Introduction: changing cultures of speed. Slow Cities 3–37. https://doi.org/10.1016/B978-0-12-815316-1.00001-0

Zipper D (2024, July 2) How e-bikes won over Europe. Bloomberg.Com. https://www.bloomberg.com/news/articles/2024-07-02/europe-s-e-bike-boom-has-lessons-for-us-cities

Daniel Piatkowski is an Associate Professor of Integrated Land Use and Transport Planning in the Department of Built Environment at OsloMet—Oslo Metropolitan University, Norway. His research is at the intersection of sustainable mobility, bicycle urbanism, and emerging transport technologies. He teaches courses in transport policy and urban planning in the Master's of Transport and Urban Planning program at OsloMet. Dr. Piatkowski is the author of the book 'Bicycle City: Riding the Bike Boom to a Brighter Future' published by Island Press.

Chaoru Lu is an Associate Professor of Smart Mobility in the Department of Built Environment at OsloMet—Oslo Metropolitan University, Norway. His research is at the intersection of sustainable mobility, transport electrification, public transport systems, and emerging transport technologies. He teaches courses in transport engineering and transport modelling in the Master's of Transport and Urban Planning program at OsloMet.

John Östh is a Full Professor of Urban Analytics in the Department of Built Environment at OsloMet—Oslo Metropolitan University, Norway and coordinating education in Transport and Urban Planning. He has written extensively on topics related to mobility, urban planning, tourism, demographics and urban economics. John has developed several scientific software of which the most downloaded (EquiPop) is used by statistical agencies and researchers in worldwide.

Chapter 11
Placemaking of Urban Green Areas: The Failed Potential of Social Interaction

Tanja Obradović, Jelena Đekić, Tamara Rađenović, Milena Dinić Branković, and Milica Igić

Abstract Urban green spaces (UGS) are key elements of placemaking, contributing to ecological balance, public health, and social cohesion. In post-socialist cities, however, their potential to foster social interaction remains largely underutilized. This paper examines the discrepancy between planned and realized functions of UGS in Niš, Serbia. The analysis combines planning documents, legal frameworks, and field observations to trace the effects of post-socialist transformations such as restitution, privatization, and deregulation. These processes have led to the loss, fragmentation, and inadequate maintenance of green areas, reducing their capacity to support community life. To illustrate these dynamics, the paper presents selected case studies: Krive Livade, Somborska–Studenička, and Vagres City, along with recent initiatives aiming to restore the social role of green spaces. The examples highlight a shift from socialist-era planning, which prioritized collective open areas, toward contemporary market-driven development that often undermines greenery and social use. Findings indicate that UGS in Niš face challenges of underuse, poor management, and spatial degradation. The paper argues that effective placemaking requires not only spatial and ecological planning but also institutional support, community participation, and coherent implementation, without which UGS cannot realize their role as spaces of interaction, identity, and resilience.

T. Obradović (✉)
Department for Inspection Affairs of Civil Engineering and Urban Planning, Ministry of Construction, Transport and Infrastructure, Niš, Serbia
e-mail: obradovic.tanja@gmail.com

J. Đekić · M. D. Branković · M. Igić
Faculty of Civil Engineering and Architecture, University of Niš, Niš, Serbia
e-mail: jelena.djekic@gaf.ni.ac.rs

M. D. Branković
e-mail: milena.dinic@gaf.ni.ac.rs

M. Igić
e-mail: milica.igic@gaf.ni.ac.rs

T. Rađenović
Faculty of Occupational Safety, University of Niš, Niš, Serbia
e-mail: tamara.radjenovic@znrfak.ni.ac.rs

J. J. Stanković et al. (eds.), *Regional and Urban Challenges and Opportunities for Data-Policy Interactions*, The Urban Book Series,
https://doi.org/10.1007/978-3-032-08592-4_11

Keywords Urban green spaces · Placemaking · Urban transformation · Social interaction

11.1 Introduction

Placemaking is a broad term encompassing different aspects of development primarily focused on public spaces, as an extension of the community (Cilliers et al. 2015). Placemaking strives to create vibrant places for people to socialize and interact, thus strengthening the sense of community. It represents a significant determinant of sustainable community development, as it entails the systematic coordination of planning, design, management, and programming efforts within spaces, aimed at fostering diverse patterns and activities across cultural, social, economic, and ecological dimensions, ultimately directed toward enhancing the overall quality of life, economic vitality, and environmental health (Turvey 2019). The importance of nature and the environment for the health and social well-being of the citizens has been especially potentiated during the times of the COVID-19 pandemic, as it has raised awareness about the consequences of human actions on the environment, but also the fact that greater participation in nature changes attitudes toward the environment (Rađenović et al. 2022; Živković et al. 2022).

Public places are the beating heart of any community, and placemaking encourages people to rethink and reinvent them together to maximize shared value (PPS 2007). It not only promotes urban design but encourages innovative patterns of use while paying attention to the social, cultural, and physical identities that shape a place and contribute to its continuous development (PPS 2007).

Placemaking is an approach in urban planning and design that focuses on the people who use the space, not just the physical structures or buildings. The arrangement of green areas is a key part of this approach, as they play a central role in activating social interaction. As part of the free, undeveloped areas that placemaking includes, it is necessary to pay special attention to the design of greenery.

Urban green spaces (UGS) play multiple roles in cities and affect the quality and livability of urban space. Urban greenery provides a wide range of ecosystem services to urban societies, such as mitigating the urban heat island effect (Price et al. 2015) and improving outdoor thermal comfort (Đekić et al. 2018a). Measurements of the surface temperature of different paving materials in the summer period showed that the temperature of the grass area was up to 20 °C lower than the temperature of the surrounding paving materials (Đekić et al. 2018b). Temperatures of materials shaded by tree tops are also significantly lower in comparison to the same material exposed to sunlight (Đekić et al. 2018b). These temperature differences are noticeable for pedestrians and affect their thermal comfort (Đekić et al. 2018a). Urban green areas also affect the mitigation of flooding (Zellner et al. 2016; Roy et al. 2012), improving air quality (Moradpour and Hosseini 2020; Jayasooriya et al. 2017), reducing stress levels, improving human health and sense of well-being (Lottrup et al. 2013; Navarrete-Hernandez and Laffan 2019), and increasing property values (Crompton and Nicholls 2022; Szczepańska et al. 2016; Melichar and Kaprová 2013).

The number of inhabitants in urban areas is constantly increasing, and at this moment, 55% of the world's population lives in cities. According to projections, the urban population will reach 68% by 2050 (UNDESA 2018). In America, the percentage of the urban population today is as much as 80% (US Census Bureau 2022; The World Bank 2022). A potential development strategy to mitigate the negative effects of population and economic growth on the environment is the circular economy and sustainable resource management (Vranjanac et al. 2023). Namely, it is important to meet the needs of current generations, without jeopardizing those of future ones. Thus, it should not be neglected the importance of aligning the citizens' needs, the accountability of local self-governments, and the national regulatory framework to fulfill the needs for public utility services, including public green areas (Vasović et al. 2023). Along with the increase in the number of inhabitants, the need for green areas in cities also grows. On the other hand, urbanization has led to an increase in the level of construction and a decrease in open areas, especially green areas. Bearing in mind the multiple benefits of green areas for ecology and sustainability, but also for the health and well-being of people in cities, recently a particularly important part of research in this context has focused on the availability, accessibility, and attractiveness of urban green and blue spaces (UGBS) for different socioeconomic groups of residents (Nesbitt et al. 2018; Kronenberg et al. 2020; Gradinaru et al. 2023).

This paper explores the under-researched area of placemaking in urban environments, focusing in particular on the role of social interaction and its (non-)realization through UGS. Green spaces in cities are much more than aesthetics. They are places for social interaction, community strengthening, and mental well-being. When effectively planned and designed, these spaces can transform cities and enrich the lives of their residents.

Although the advantages of green spaces in promoting social interaction and well-being are recognized, in Serbia there is still a large gap between the planned and realized results of such spaces. There are few works dealing with impact analysis, as well as identifying the causes that would ultimately lead to an improvement in the situation in this area. This paper critically analyzes the factors that influence the failure to achieve the goals of placemaking through the realization of green spaces. It explores the challenges and omissions in the planning and use of green spaces that significantly affect the community.

Since 1990, the countries of Central and Eastern Europe (CEE) have gone through a process of post-socialist transformation. In the conditions of dramatic multi-layered post-socialist changes, including changes in urban and housing policy, as well as changes in the role of urban planning and changes in the legal status and ownership structure of urban construction land, the preservation of inherited green spaces and the planning of new green spaces become serious challenges for planners and public authorities. A large part of green spaces in post-socialist cities is not protected as parks, forests, or other forms of green spaces, although such spaces and their functional connections and interrelationships provide a wide range of services to city residents (Kronenberg et al. 2020). Therefore, the question arises as to whether placemaking is possible under the newly created conditions and in what way it can be implemented.

Loss of public open space (POS), particularly loss of green space in favor of building construction, is especially noticeable in the post-socialist cities due to the rapid loss of greenery for a short time after the conversion from public to private ownership over land. The planning of UGS is an integral part of urban planning, so changes in the planning and development of UGS can be observed through urban planning documents. However, since urban planning documents are rarely fully implemented, a true insight into the state of green spaces can only be obtained through field research and comparison with planning documents, to spot deficiencies in planning and implementation.

The paper aims to answer two main questions:

(1) Whether and in what way post-socialist transformations affected the status and development of public open and green areas in cities, especially green spaces in areas of multi-family housing, as the dominant type of housing in cities in the socialist and post-socialist period
(2) What are the perspectives of placemaking through green areas in changed conditions?

On the example of the City of Niš, the socialist and post-socialist context of the development of green areas in areas of multi-family housing is analyzed. In order to better understand the development and status of green areas, the planning and legal framework is analyzed, as well as the situation in the area as a result of the (non-) implementation of these documents. Comparison of multi-family housing areas from different periods provides insight into the potential for the development of green areas and their improvement for social interactions.

11.2 Post-socialist Context

In the period between 1989 and 1992, communist regimes fell throughout the region, which was followed by a deep political and economic crisis, and in some countries by ethnic wars and the disintegration of states. The end of state socialism marks the beginning of reforms leading to market economies and democratic governance, hence the term "transition" commonly used for the post-socialist period (Hirt 2015; Sýkora 1994). In the transition period, most of the countries of the region recorded an increase in unemployment exceeding a quarter of the population, an explosion in the inflation rate reaching double and triple-digit values, the collapse of state-guaranteed services (social security, unemployment benefits, state health care, etc.), a drastic drop in the standard of living of the majority of the population, and a sharp increase in social stratification (Hirt 2015).

After the fall of the authoritarian regimes, in the 1990s, a democratic political system was established throughout the region, and the decentralization of power toward the local level, that is, the strengthening of local self-government units (LSGU). Decentralization has led to the transfer of numerous responsibilities from the central to the local level, thus presenting local authorities with challenges related to the provision of adequate infrastructure, transport management, and land use planning (Tsenkova 2006), but, without providing an adequate budget that would enable the provision of quality services and dealing with the problems of post-socialist cities. Municipal finances were still largely dependent on transfers from the central budget, which did not keep pace with the increase in local responsibilities. Local authorities have mostly become responsible for urban planning and management, providing services and maintaining technical infrastructure, and providing social services. In most cases, municipalities have acquired ownership of public utility companies (water supply and sewerage companies, central heating systems) and public spaces for rental (Tsenkova 2006; Hegedüs et al. 1993). However, the central government continues to influence local urban development, through the adoption of legal regulation of market processes, control of financial relations between local governments, and the transfer of money from the central budget to local ones (Backović 2005).

Multilevel transformations were also reflected in changes in urban and housing policy, as well as in urban planning changes. Privatization of government housing at below-market prices in many countries in the region has led to the formation of a society of poor or economically paralyzed homeowners (Mandič 2010; Tsenkova and Polanska 2014). In this way, the responsibility for the maintenance of buildings was transferred from the state, that is, state enterprises, to the owners, who did not have enough funds and knowledge to manage the property, which led to the degradation of the inherited housing stock.

The status of POS, including the spaces of the inherited large housing estates (LHE), depended on the transition path. In some countries, such as Hungary, the restitution process was implemented very early through symbolic compensation in the form of coupons or agricultural vouchers of limited value, which were awarded to pre-war owners (Marcuse 1996). This way, POS, including those in inherited LHEs, remained the property of the municipality and was not fragmented by returning the nationalized land to its owners.

On the other hand, most countries, including the former Socialist Federal Republic of Yugoslavia (SFRY) republics, chose to implement "in kind" restitution. The restitution laws enacted in the 1990s made the POS in inherited LHEs legally eligible for private development. The undeveloped land, previously intended for the construction of supporting facilities for housing (commerce, administration, and public services), but also the POS of LHEs, became a suitable land for post-socialist development. Former playgrounds and parking spaces became privately owned small businesses or individual or collective residential buildings (Marin et al. 2023).

11.3 The City of Niš Case Study

Green spaces in cities are a huge potential and driving force for placemaking. They are a place for rest, relaxation, recreation, and connection with nature. Furthermore, they serve as places for joint community events, cultural activities, and social gatherings, fostering a sense of belonging and identity among residents. From parks, squares, and gardens to waterfront promenades, these green spaces are integral components of urban structures, enriching the quality of life of city residents.

The City of Niš and its surroundings, although recognized as the center of biodiversity of the Balkans, does not use or improve its natural potential.

11.3.1 Niš the City of Change

The post-war period in the development of the City of Niš was marked by great demographic growth and spatial expansion. In the period from 1945 to 1980, the number of inhabitants and apartments in Niš increased more than four times, while the effectively constructed city space increased more than six times (Krstić and Medvedev 1983). Niš was the administrative and economic center of the wider area and one of the Yugoslav cities with the most dynamic urbanization after the Second World War. The development of industry, which required a large number of employees, created opportunities for employment and the influx of population from rural areas. The spatial migration of the population from the countryside to the city led to the concentration of the population in the city area and its immediate surroundings (about 82% of the total population) and to deconcentration in the rest of the gravitational area of the city (Krstić and Medvedev 1983).

At the time of the massive influx of population, there was no appropriate urbanization policy that would cover the wider area of the city and direct the mechanical influx of population. The same as in other socialist countries, in the case of Niš, the increase in the number of inhabitants and the dynamic economic development was not accompanied by adequate communal and housing construction. The disproportional development of the economy and the communal and housing sector on the one hand, and the mechanical influx of the population on the other, in conditions of disrespect and non-existence of urban norms, led to the appearance of massive illegal housing construction on the outskirts, and the expansion of the city. During the 1960s and 1970s, the population of the City of Niš doubled, which was accompanied by the development of mass housing construction to provide housing for newly arrived residents. The 1970s are characterized by increasingly significant concentrated construction of residential buildings in the social sector on wider, empty spaces, which are specially equipped for the needs of mass housing construction.

The period of the 1990s in all CEE countries was marked by the industry decline, the economic crisis, and a large increase unemployment rate. In Serbia, that is Federal Republic of Yugoslavia (FRY), the situation is particularly dramatic due to the collapse of the federation, war conflicts, and long-term international isolation. The 1990s in Serbia were marked by a sharp economic decline and a deep socioeconomic crisis. Niš, which was one of the leading industrial centers in the former SFRY, failed to reposition itself in the years after the 1990s and experienced a dramatic economic decline and population stagnation (Vasilevska et al. 2015). The unemployment rate reached 39.4%, and the average income was around €150 in the years around 2005, which was far below the national average (Vasilevska et al. 2015).

Under the influence of the economic decline and reduced allocation of the budget for housing development (approximately 0.1% of the city budget from 2000 to 2005), the housing sector in Niš experienced regressive changes (Vasilevska et al. 2015). The most important changes since 1990 relate to the absence of significant public investment in housing development and the complete lack of investment in rental and social housing and maintenance, as well as the double reduction in housing production and the closure of leading public construction companies. In 2003, that number was reduced to around 1.8 apartments per 1,000 inhabitants.

After the political changes in 2000, the gradual recovery of the economy began. The gross domestic product increased in 2003 by 10.8% compared to 2000, but compared to 1990 it was still at a very low level (52.8%) (Vasilevska et al. 2015). A gradual recovery is also noticeable in the housing sector. The total number of completed apartments in 2012 was 529, i.e., 2 apartments per 1,000 inhabitants (Action Plan for Sustainable Development of the City of Niš 2015–2020), which indicates that two decades after the start of the transition, residential construction in Niš is on the slight rise, but still at a lower level compared to the pre-transition period. The construction intensity has been particularly noticeable in the last 10–15 years, as evidenced by the statistics of issuing building permits. To achieve the construction intensity, planning documents are prepared and adopted according to the methodology of defining the physical structure. Although the intensity has increased, which contributes to economic development, new practices in planning, resulting from the regulations after 2003, are characterized by the process of deregulation, especially in the part of conditioning in terms of physical and design characteristics of buildings, preservation of ambient characteristics and characteristics of the climate, clear typologies and types construction, and the formation of harmonious urban units, pleasant for people's lives (Obradović and Polić 2024).

11.3.2 Conceptualization and Structure of Green Spaces in Multi-family Housing Areas in the Socialist Period

The planning of socialist LHEs, based on modernist principles and the unit of neighborhood or micro-region as the basic unit of planning, meant above all: (1) ensuring a sufficient amount of available greenery; (2) positioning of buildings so that apartments receive enough sunlight and (3) traffic planning to improve pedestrian safety

(Treija and Bratuškins 2019; Tuvikene 2019). Modernist principles shaped socialist housing estates to such an extent that it is often stated that the modernist concept of the "tower in the park" found its purest expression precisely in the microrayons of the Soviet Union (Grava 2007). Although in the construction of residential settlements, the emphasis was on efficiency and quantity, at the same time it was insisted on the quality of housing. Namely, the organization of settlements with free-standing buildings and large green areas was supposed to ensure optimal microclimatic and sanitary conditions for housing, as well as recreation areas near the apartment. On the other hand, state control over urban construction land allowed planning and disposal of space with less economizing (Szelényi 1996), which allowed socialist neighborhoods to be exceptionally well supplied with parks and other POS (Hirt and Kovachev 2006). The importance of greenery was reflected in the creation of a favorable microclimate, the provision of hygienic conditions and space for relaxation and recreation, and the formation of an urban landscape; whereas the share of green areas was usually around 40–45% of the total area of the settlement (Treija and Bratuškins 2019).

Free spaces in residential areas, in addition to green and recreational areas, are also planned with facilities for common (public) use. It is planned that some activities from the private domain of life will be moved to common (cooperative, communal) spaces; thus, the border between the private and public domains of life will gradually be lost to merge them into one. Open and pedestrian areas of newly built settlements, with the potential for socialization and gathering, were seen as the highest reach of humanization and were given great importance (Gulin Zrnić 2017).

In addition to ideological influences, the size of the free space between buildings was also determined by construction technology. The use of large cranes for the assembly of prefabricated elements required space for maneuvering, which also affected the dimensioning of the distance between buildings and the size of open spaces; in some cases, the projects were designed to enable the construction of two buildings using one crane position—"crane track principle" (Bakay 2012). Large free and green spaces are still one of the main characteristics of socialist housing estates today and most of these housing estates are considered very green compared to other parts of cities. Free-standing buildings surrounded by abundant greenery are often mentioned in the descriptions of these settlements (Treija and Bratuškins 2019; Benkő 2015).

Socialist housing estates in Yugoslavia share the characteristics of housing estates in neighboring countries, but they are also distinguished by certain specificities. The liberal nature and specifics of the Yugoslav self-governing socialist system resulted in a better quality of life and richer architecture in LHEs compared to other socialist countries (Hirt 2008).

The period 1968–1978 in Niš is characterized by increasingly significant concentrated construction of residential buildings in the social sector on wider, empty spaces, which are specially equipped for the needs of mass housing construction. The planning of areas for the construction of large residential settlements was possible since the Law on the Nationalization of Rental Buildings and Construction Land from 1958 (Official Gazette of FNRY, no. 52/58), in addition to rental residential and commercial buildings, also nationalized construction land in cities and urban settlements. The Constitution of SFRY from 1974 (Official Gazette of SFRY, no. 9/74) guaranteed farmers the right to own arable agricultural land in an area of no more than ten hectares per household. The Constitution also stipulates that there can be no property rights for land in cities and settlements of urban character and other areas intended for residential and other complex construction, which, in accordance with the conditions and procedures established by law, are determined by the municipality. In this way, conditions were created for the use of land for the construction of housing estates and other state projects, without taking into account the ownership of land and the price of urban construction land.

The structure of urban planning in the socialist period consisted of a general urban plan (GUP) and detailed urban plans (DUPs), which partially developed parts of the city for the needs of industrial development, the arrangement of social centers, sports-recreational and green areas, zones of collective or individual housing. Although the General Plan of Niš from 1953, followed by the GUP of Niš and the Spa of Niš from 1973, were in force during the period of the most significant residential construction in Niš, most of the residential construction was carried out according to DUPs that were partially adopted for housing construction needs. The decision to implement the GUP was made in 1978, and in the meantime, the first amendment was already made in 1974, just one year after its adoption. Mass illegal housing construction, especially in areas that were not planned for housing (public protective greenery and forests), requests for the construction of economic capacities in an area not intended for that purpose, and inconsistency of traffic solutions, required frequent changes to the GUP. Thus, in ten years after its adoption, as many as eight amendments to the GUP were adopted (Medvedev 2018).

In the conditions of constant revisions and changes of GUPs, the development of the city is guided by DUPs, so the majority of residential areas were built according to the solutions of DUPs (Krive livade—Krivi vir, Trg radnički saveta, Braća Tasković, Ćele Kula, Delijski vis, Crvena zvezda, Čarnojevića, Bubanj, Somborska, Durlan, etc.). In addition to the detailed elaboration of traffic and the disposition of residential buildings, the DUPs also contained a landscape solution of free space with the disposition of sports fields and children's playgrounds, low, medium, and high vegetation.

Following the Law on the Establishment and Maintenance of Green Areas (Official Gazette of the SR of Serbia, No. 8/72), the Program for the Establishment and Maintenance of Green Areas in Niš and the Spa of Niš was drawn up. The program lists all existing green areas, classified into several categories according to the law:

(1) public green areas, parks, squares, rows of trees, lawns, greenery along roads, forest parks, etc.;
(2) protective belts of settlements and industrial zones, agricultural protection belts, greenery along the banks of rivers, and other water bodies;
(3) green areas of special purpose and limited utility: botanical gardens, arboretums, memorial parks, archeological parks;
(4) green areas of work and other organizations;

(5) green areas for sports and recreation (playgrounds for children, sports yards and practice areas, sports fields and recreational areas);
(6) greenery of yards, gardens, weekend settlements, camping areas, picnic areas, etc.

Today, such categorization is not established by law in Serbia, nor is the maintenance program linked to planning documents. Some countries of the former SFRY still have a law defining the categorization and management of green areas. Such an example is the law from the Republic of North Macedonia, the Law for Urban Greenery (Official Gazette of the Republic of Macedonia, No. 11/2018).

During the 1960s, certain planning or project documentation (detailed urban solutions, landscaping, etc.) had already been drawn up for most of the parks in the city, while block areas, free spaces of residential areas, had yet to be developed. In the conclusion, it is stated that despite the detailly elaborated landscape solutions, the blocks were left without greenery that needs to be arranged (Fig. 11.1).

Fig. 11.1 Green areas of the inherited socialist housing estates in Niš that have not undergone transformations. *Source* https://free-map.org/

11.3.3 Green Spaces of High-Rise Residential Areas in Post-socialism

11.3.3.1 Changes in Urban Planning and the Status of Urban Construction Land

The transition period is characterized by the return of ownership rights to real estate and land, through privatization and restitution processes, which took place at different rates in the countries of the region, depending on the transition path. In accordance with the trend of privatization of social housing in Serbia, the housing stock in Niš was almost completely privatized during the 1990s (98.7%) (Study on the Development of the Local Economy of the City of Niš 2005). One of the key catalysts for the dynamics of urban functions, especially urban green areas, was the legislative change that allowed the restitution of previously nationalized land. As stated by Hirt and Stanilov (2007), the roots of the process of green space disappearance can be traced to the unfortunate interpretation of land restitution laws, which allowed the restitution of land as long as there were no buildings on the ground.

In the initial phase of the transition, the Constitution of the Republic of Serbia from 1990 (Official Gazette of RS no. 1/90) determines that urban construction land, as well as natural resources and goods of general use, as goods of general interest, are state or social property. The Law on Construction Land from 1990 (Official Gazette of the SR of Serbia, no. 23/1990) confirms that there can be no ownership right on city construction land, but according to the Constitution, the right of use can be acquired. Soon after, in 1992, the Constitution of the FRY (Official Gazette of FRY, no. 1/92) determined that certain goods in general use and urban construction land could be privately owned and in other forms of ownership. The Law on Construction Land (Official Gazette of RS, no. 44/1995), which was adopted in 1995, states that construction land is in circulation, unless otherwise specified by law, but also that urban construction land is state property and cannot be alienated from state property. Only at the beginning of the 2000s, with the adoption of the Law on Planning and Construction in 2003 (Official Gazette of RS, no. 47/2003), construction land was divided into: (1) public construction land, which cannot be alienated from state property; and (2) other construction land, which can be in all forms of ownership and is in circulation.

By amending the Law on Planning and Construction from 2009 (Official Gazette of RS, no. 72/2009, 81/2009, 64/2010 (Decision of Constitutional Court), 24/2011, 121/2012, 42/2013 (Decision of Constitutional Court), 50/2013 (Decision of Constitutional Court), 54/2013 (Decision of Constitutional Court), 98/2013 (Decision of Constitutional Court), 132/2014, 145/2014, 83/2018, 31/2019, 37/2019 (other law), 9/2020, 52/2021), construction land is defined as follows:

- Construction land is land that is designated by law or a planning document for the construction and use of buildings, as well as land on which buildings are built in accordance with the law.

- Construction land can be in all forms of ownership. The Republic of Serbia, an autonomous province, i.e., LSGU has the right of ownership on construction land in public ownership.
- The construction land is in circulation. Construction land in public ownership is in circulation, under the conditions prescribed by this law and other regulations.

In the initial phase of the transition, construction land is in circulation, while urban construction land is state property and cannot be alienated. This status of the land has influenced the fact that residential construction begins to develop in an area that, according to urban plans, is not designated as urban construction land because it is land in circulation. Under these conditions, housing construction begins at the Somborska-Studenička location. Later changes in the status of urban construction land, as well as the progress of the restitution process, will affect the appearance of infill construction on green areas of inherited multi-family residential areas, and the consequent fragmentation and reduction of POS and green spaces in these neighborhoods as well.

Urban planning includes three levels of planning: GUP, general regulation plan (GRP), and detailed regulation plan (DRP).

The analysis of general urban plans for the city of Niš adopted after 1990 (GP1995, GUP2010) showed that they generally recognize similar categories of greenery (Table 11.1). In addition to greenery as a separate land use, the plans foresee greenery as part of other land uses (health, education, housing, business, etc.).

There is a notable absence of established criteria and a lack of cohesive planning concerning urban green spaces. The sole standard mentioned in these plans pertains to the surface area of green spaces per capita, while the size of green areas, their function, and proximity to residential areas are entirely overlooked. Furthermore, both plans exhibit shortcomings in terms of not including provisions for new public green spaces and featuring low percentages of greenery within other land use categories, particularly housing (Đekić et al. 2017).

The greenery of inherited residential areas is not registered in the plan, but entire settlements—built-up and free spaces—are designated as housing, which, along with the change in the possibility of turnover of city construction land and "in kind"

Table 11.1 Comparative review of urban greenery in GP1995 and GUP2010 (Đekić et al. 2017)

GP 1995	GUP 2010
Greenery in the buildable area	
City greenery	**Green and open spaces**
Parks	Recreational area
Park forests	Park greenery
Protective greenery in the buildable area	Park forests
Other greenery	Protective greenery (mainly other land)
Greenery outside the buildable area	
Protective greenery outside the buildable area	Forests and forest land

restitution, opened the possibility for additional construction in the area of high-density residential areas.

11.3.3.2 Development of Housing and Green Areas

Although the GUPs showed an increase in green areas in the city of Niš, no new green areas were actually planned in the city core, and the increase was achieved by increasing the scope of the plans and including the surrounding agricultural land in the scope of the GUP.

The volume and characteristics of housing construction in the post-socialist period depended on the availability of land for construction and private capital (Đekić and Vasilevska 2021). During the early stages of the transition, the private sector was limited and not fully developed. Additionally, there was little or no available building land in central city areas due to unfinished restitution and unresolved property rights issues. This led to the expansion of multi-story housing extensions within the inner city (construction without a building plot) and new developments cropping up on the city's outskirts. Multi-story housing extensions were the dominant form of residential construction in the 1990s in Niš. In addition to having a significant impact on the increase in the number of inhabitants, previous research has shown that building upgrades also affect the thermal comfort of users of open spaces (Dinic-Brankovic et al. 2022). In the subsequent phase, there was notable growth in the private sector, along with improved access to construction land in the central urban zone, which enabled more intensive construction in the central city areas, including areas of inherited socialist housing estates, as well as larger-scale construction (Đekić and Vasilevska 2021).

Since residential construction has been developing intensively in recent years, and there is no development of new public green areas, green areas within housing can make a significant difference in the quality of residents in cities. In addition, a large share of the total housing in the city of Niš is still made up of inherited multi-family housing from the socialist period, characterized by large areas of greenery. Therefore, the development of green areas in the post-socialist period must be viewed in two directions: (1) green areas in inherited areas of multi-family housing; (2) green areas within the new multi-family housing.

11.3.4 Examples

At the heart of placemaking urban green spaces lies social interaction—spontaneous encounters, sharing experiences, and a sense of community that contribute to a rich social life. Social interaction at these locations encompasses a wide range of activities, from sporting activities to casual walks and conversations that connect people. These interpersonal relationships not only strengthen social bonds but also foster a sense of collective responsibility and care for public spaces, which contributes to their vitality and sustainability. The absence of planning for such spaces, or their planning but non-realization in the space, can lead to a reaction from the community whose basic need it is.

In continuation, the state and potential of the green areas of residential areas on the territory of the City of Niš, built in the socialist and post-socialist periods, are analyzed: (1) Krive Livade—a typical example of a socialist residential area; (2) Somborska-Studenička—a settlement that began to develop in the 2000s based on plans from the socialist period in post-socialist conditions; (3) Vagres City—a settlement entirely planned and built in the post-socialist period. Transitional changes that began in the 1990s, in Serbia, and most of the former republics of SFRY almost a decade later are still not fully completed and continue to transform existing spaces and shape the conditions for the development of new green areas.

11.3.4.1 Krive Livade—Green Spaces of Inherited Areas of Multi-family Housing

The development of the Krive livade settlement began in the 1970s at the time of the most intensive residential construction in Niš. The DUP for the neighborhood level clearly defined the uses of the areas (housing, public functions: school, kindergarten, local community), traffic, dimensions and disposition of buildings, as well as detailed arrangement of free space (greenery, recreation, children's playgrounds). Residential buildings and inter-block spaces were fully realized according to plan, while the construction of commercial and public facilities (store, school, local community building) was largely absent. The inherited area was characterized by large open spaces, further enlarged due to undeveloped planned contents.

It was a period of great growth and economic reforms (1961–1974), also marked as the period of the beginning of spatial planning, which was followed by institutional and legislative reforms (Obradović and Mitković 2012).

Since political, institutional, economic, and other changes in the post-socialist period took place at different speeds, urban planning did not transform quickly enough following the new conditions but still rested on socialist foundations. According to the urban plans from the 1990s, the area of the Krive livade settlement was mostly designated as housing with a density of 200–250 inh/ha (Fig. 11.2a). Later plans from the 2000s also built the inter-block spaces of the Krive livada area, marked as high-density housing in the city area (Fig. 11.2b).

POS in the settlement are not designated as green areas, nor are the construction rules for green areas applied to them, but on the contrary, the area is treated as "high-density housing," which creates the possibility of further construction on the free space between the buildings. This area is positioned as a secondary city center in the post-socialist period. Due to changing market conditions in the late 1990s, this plan was changed several times. Changes to the plan enabled the construction of shopping centers and affected the system of city centers and the urban fabric (Dinic-Brankovic et al. 2018). The position of the settlement along a busy road that provides an excellent connection with the city center and other parts of the city, proximity to the city center,

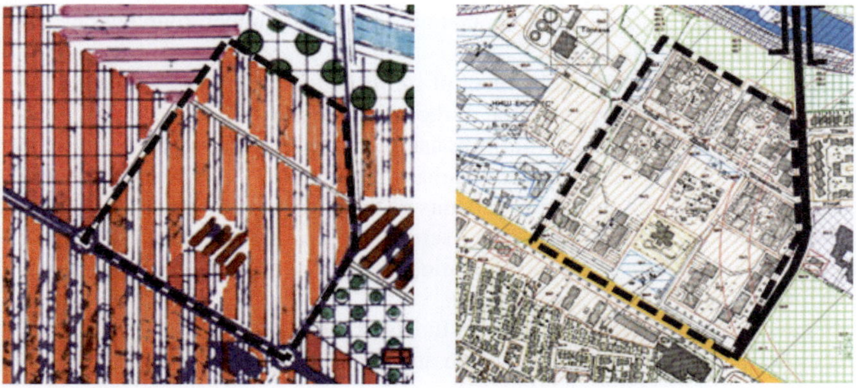

Urban housing density 200-250 inh/ha
a)

High-density housing in the urban area
b)

Fig. 11.2 Land use in inherited areas of multi-family housing according to GUPs. **a** GP Niš 1995–2010; **b** GUP Niš 2010–2025. *Source* Authors (Đekić 2022)

infrastructural facilities, and the development of commercial activities, made Krive livade attractive for new investors in the post-socialist period.

At the same time, the beginning of restitution and changes in the law by which land in all forms of ownership is put on the market enable private investors to acquire land for construction. As is the case in other countries that have chosen "in kind" restitution, every piece of free land that is not occupied by buildings, i.e., brought to a purpose, becomes subject to restitution.

New commercial and residential developments contributed to the restoration of the social structure and regeneration of the area, on the one hand, while, on the other hand, they manifested their negative effects in the form of spatial fragmentation and degradation of green areas. Changes in green areas are manifested as: (1) quantitative—reduction of green areas and (2) qualitative—fragmentation, change of purpose and character of areas.

About 1.5 ha of POS, including space intended for the construction of an elementary school, was taken up by residential infill construction, which represents a tenth of the total area of multi-family housing Krive livade. The construction of new residential buildings implies the provision of public or private access roads to the buildings, which leads to additional fragmentation of POS and a significant increase in traffic inside the block, where pedestrians and cars compete for space (Vasilevska et al. 2014). Additional residential construction led to an increase in the number of residents and the number of cars, and consequently to an increased need for parking space. In addition, the development of commercial activities further increased the need for parking space. The existing parking spaces, planned according to socialist norms, do not meet today's parking needs; the increase in parking space usually takes place at the expense of free (green) areas. The number of parking spaces in the area has increased almost three times, not counting parking in underground garages next to newly built buildings.

Figure 11.3 shows the development of traffic in the multi-family housing Krive livade and the gradual fragmentation of the area.

Green areas within the new (infill) housing construction are represented by 10% of the building plot area. These green areas are very rarely complete; they usually appear as shredded and scattered remnants of the space during construction, which have minimal or no use value. When setting the price per square meter, private investors refer to public green and recreational spaces in the surrounding area, which are being degraded by new construction.

Figure 11.4 shows the green areas of the inherited housing and the free spaces of the new infill construction without green areas.

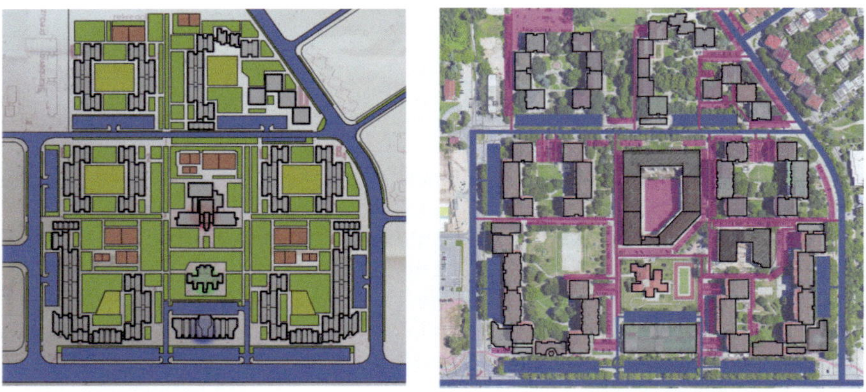

a) b)

Fig. 11.3 Traffic changes from settlement planning (multi-family housing) to today: **a** DUP Krive livade 1978; **b** current state (blue—planned traffic; magenta—newly introduced traffic and parking). *Source* Authors

a) b)

Fig. 11.4 Open spaces and greenery within: **a** inherited housing, and **b** infill construction. *Source* Authors

Table 11.2 Comparison of urban planning parameters in the socialist period and after post-socialist transformations

		Socialist period	Post-socialist transformation
Number of inhabitants		9.547	13.734
Population density	PD (n.u.)	613 inh/ha	881 inh/ha
	PD (h.b.)	728 inh/ha	1048 inh/ha
Number of parking places		664	1,400 (+350 in underground garages)
Green and open spaces		Around 7.3 ha 73% of the area intended for housing	Around 4.8 ha 48% of the area intended for housing

PD (n.u.)—population density of the entire neighborhood unit
PD (h.b.)—population density of housing blocks
Source Authors

According to the DUP of Krive livade in 1978, residential and commercial buildings occupied about 1/3 of the space intended for living, while landscaped free and green areas occupied about 2/3, i.e., >70% of the space intended for living. Today, free and green areas of the settlement make up about 48.3% of the area intended for housing (Table 11.2).

11.3.4.2 Somborska—Green Spaces of the Multi-family Residential Area in the Early 2000s

The development of multi-family housing on undeveloped land along Somborski Boulevard began in the early 2000s in the period of the "real" transition or the beginning of the transition for the second time in Serbia (Uvalić 2010). The Somborska-Studenička area was planned for the future expansion of multi-family housing back in the 1980s. Planning based on the neighborhood unit (local community) provided, in addition to housing, facilities characteristic of the local community level: school, kindergarten, settlement center, and park areas, as in the case of the Kriva Livada area. The development of the area, however, began in significantly changed conditions: changed property regime on city construction land, change of actors in housing construction, the emergence of illegal construction, weak enforcement control, and frequent changes of urban plans. The GUP of Niš 2010–2025 has undergone four changes since its adoption, while certain GRPs, which define the rules of arrangement and construction, have been changed six times.

The appearance of the Somborska-Studenička area is the result of arbitrary interpretation and neglect of planning solutions to achieve maximum profit, that is, investor urbanism that is supported by planning documents. The DRP from 2004, still in the drafting phase, had to adapt the planning solutions to the existing illegal construction. Changes to the DRP followed soon, to legalize the newly created deviations in construction.

In addition, the level of detail of the planning documents influenced and still influences placemaking. The DUPs, which were used in the socialist period to develop areas of collective housing construction, showed in detail the position and number of floors of the buildings, traffic, and parking in the settlement, the arrangement of free spaces, planting material, etc.; i.e., they gave the final appearance of the settlement. On the other hand, DRPs in the post-socialist period, define guiding construction rules at the building lot level (maximum number of floors, minimum distance between buildings, etc.), without showing the planned appearance of the area. This approach in planning enabled the development of the area according to the "plot by plot" principle, which resulted in the present appearance of the area. Multi-family residential buildings intended exclusively for the market were built on individual plots without considering the wider context, which led to the fragmentation of space and the absence of spatial integrity (Vasilevska et al. 2014).

The greenery in the area of the Somborska-Studenička settlement is planned as: (1) public greenery and (2) accompanying greenery, i.e., greenery on land for other purposes (housing, center of the settlement with housing). Public greenery is provided in the form of rows of trees and strips of ground floor greenery along certain streets, without planned park areas.

Unlike the intensive development of private residential construction, the development of infrastructure, which is the responsibility of the public sector, is delayed due to a lack of finance. The construction of the street started only 15 years later without the originally planned linear parterre greenery and rows of trees, i.e., without public green areas.

Regarding the greenery on the land for other purposes, the plan gives the following recommendations: "Within the arrangement of green and recreational areas in residential areas, envisage the construction of children's playgrounds, as well as places for quiet rest surrounded by tall deciduous and coniferous species with a phytocidal effect of high bio-aesthetic values and weeding of surfaces" (DRP of the "Somborska-centar" settlement ("Official Gazette of the City of Niš," no. 106/04)). This formulation is a legacy of the socialist way of thinking about space, and its implementation would be possible in the socialist context of state housing construction on state-owned land; however, in the changed post-socialist conditions, this formulation is completely inapplicable. Namely, the plan provides recommendations for arranging free and green areas in "residential zones," while not defining the obligations of investors at the level of individual construction plots. As the plan does not define the minimum share of green areas on the construction plot, green areas appear only where there is no possibility of further construction or use of the area for parking—that is, as "remaining space during construction." The formation and arrangement of free and green areas according to the recommendations of the plan are left to a future investor, according to the principle "not in my yard."

The mandatory percentage of greenery at the level of the construction plot was introduced only with the adoption of the GUP Niš 2010–2025 ("Official Gazette of the City of Niš" no. 43/11), i.e., the GRP of the SM Pantelej—Phase I ("Official Gazette of the City of Nis" No. 102/2012), when most of the existing housing has already been built. Green areas appear in places in the form of narrow strips next to the parking lot or pieces of greenery of several square meters (Figs. 11.5 and 11.6). The minimum percentage of free and green areas is set collectively, i.e., no minimum green areas are defined, nor are "free" areas clearly defined. The result of this formulation is the representation of green areas with 0% and the use of almost 100% of free areas for parking. The planning documents oblige all investors to provide parking and open green areas within their plots. Since the parking lot is strictly defined, and the open space is only formal, in practice the open space is completely neglected (Vasilevska et al. 2014). In buildings where parking is arranged in an underground garage, the dimensions of the underground garage are usually larger than the dimensions of the building itself, so that with the access roads, they occupy almost the entire area of the plot, and the roof of the garage is also used for parking. In the case when there are no garages, the free areas are used entirely for parking.

The larger green area in the built-up part of the settlement consists of cadastral plots that have not been converted into construction land, but are still used as agricultural land. After the acquisition of the land by the investor, preparatory works for the construction of buildings were recently started on these plots as well, which caused protests from the residents of the settlement. Slogans displayed on the buildings indicate the problems of settlement development (investor urbanism, lack of green and recreational areas, etc.) and express dissatisfaction with living conditions in the settlement (Fig. 11.7).

After negotiations between the city authorities and investors, a solution was reached, which was accepted by both parties, to exchange the plots that were owned

a) b)

Fig. 11.5 **a** Appearance of the built-up part of the settlement. *Source* https://free-map.org/; **b** positions of green areas in the settlement. *Source* Authors

Fig. 11.6 Green areas in the settlement. *Source* Authors

Fig. 11.7 Slogans on buildings against planned construction. *Source* Authors

by the City of Nis with two plots owned by investors on which building permits were issued. Thus, the space was given a public purpose in public ownership, on one plot of landscaped greenery and a children's playground, and on the other a kindergarten that is currently under construction. In this way, a problem that has not been systematically solved until today, and which originates from the disconnection of planning documents with the implementation of planning solutions, was only temporarily solved.

11.3.4.3 Example 3: Vagres City

The residential complex "Vagres City" was developed on a greenfield location in the wider city center, on a location intended for housing. The development of the complex began in 2016 in accordance with the valid urban plans. The residential complex is located in an attractive location near the City Park Čair, well connected to the city center. Since it is a plot privately owned by an investor and there were no restrictions on construction originating from existing buildings on the plot or in the surrounding area, there were ideal conditions for the development of a high-quality and functional residential complex.

Unlike the previous example, which is characterized by illegal construction and legalization through the amendment of planning documents, in this case, the entire procedure of creating planning documents that elaborates the location was carried out according to the valid procedure. However, the detailed planning of the location was entrusted to a private company, which resulted in planning solutions in accordance with the investor's request, without considering the interests of future residents of the area.

According to the valid GRP for this area, greenery is planned as park greenery, protective greenery, forest greenery, and greenery within the profile of traffic roads. According to the same plan, the subject area is intended for high-density housing in the urban area. When talking about greenery within a housing, it is defined in two ways: (1) as a mandatory percentage of greenery on the construction plot for various purposes; and (2) in the case of "block greenery," according to the number of inhabitants, namely 8 m^2/inh in the reconstruction of the green area of the block, that is, 16 m^2/inh in the formation of new block greenery.

The "Vagres City" complex is located in the residential area of large densities in the city area, for which a maximum floor height of $P + 6$, an occupancy index of the construction plot up to 70%, and a minimum percentage of green areas on the plot of 10% are foreseen. Since the plan does not clearly define what is meant by the "formation of new block greenery," i.e., in which cases the norm of 16 m^2/inh is applied, the greenery is dimensioned in relation to the surface of the construction plot. The text of the DRP, which was adopted based on this GRP, foresees 11% of greenery. However, if one takes into account the fact that the reference area of the complex, on the basis of which the percentage of greenery is calculated, is reduced by excluding areas for traffic and other purposes, it is clear that the final percentage of greenery is far below the otherwise low values.

For the sake of comparison, in a residential complex with 55,000 m^2 of living space, i.e., about 550 apartments, which includes "Vagres City" (https://vagres.rs/en/vagres-city/) according to the norm of 20 m^2 of living space per inhabitant, the approximate number of inhabitants would amount to about 2,750. For this number of inhabitants, the area of "block greenery" (16 m^2/inh) should be 44,000 m^2 (4.4 ha), i.e., more than half of the area intended for construction. The area of greenery in relation to the area of the construction plot (10% of the construction plot) should be 0.82 ha.

By looking at the existing condition, it can be seen that green areas appear only as remnants of space during design and construction (Figs. 11.8 and 11.9). The GRP also states that during the construction of the parking lot, one tree must be provided for each two parking spaces, which was also not respected.

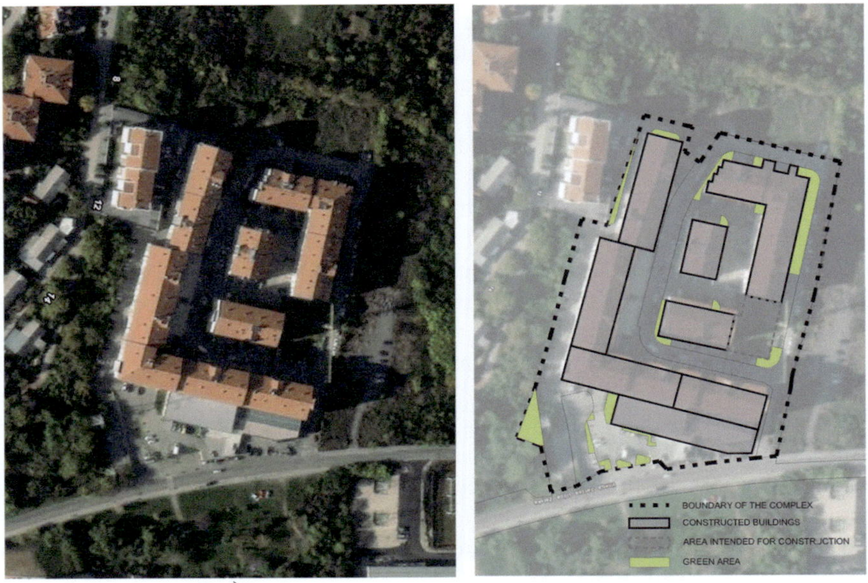

a) b)

Fig. 11.8 Vagres City. **a** Appearance of the complex. *Source* https://free-map.org/; **b** positions of green areas. *Source* Authors

Fig. 11.9 Vagres City. *Source* Vagres Archive, with permission

This is an example that both the preparation of the planning document and the implementation of planning solutions were initiated by one actor. It indicates the possibility of the planned solutions being implemented. However, it is evident that even in such cases, free and green areas are not taken into account, that is, everything that can initiate social interaction. The space that is not built in this example is intended for parking. This example shows that although there are no well-planned and implemented public spaces, it still does not affect the sale of apartments. This further leads to the conclusion that the market largely dictates how planning documents are prepared and realized, and that an additional level of increase in standards and awareness of the importance of such spaces is needed, to expect a better situation in the built space.

11.3.4.4 New Initiatives and Plans

In the last four years, there have been initiatives at the level of the City of Niš, which indicate a certain improvement in the area of green planning. In the individual GRPs, Niš introduced binding urban planning parameters that require a higher percentage of greenery and free areas. In the procedures for amending the planning documents for the central area (Third Amendments and Supplements to the GRP of the CM of Medijana), as well as for part of the Crveni Krst municipality (Third Amendments and Supplements to the GRP of the CM of Crveni Krst—Phase I), when defining the building rules, the mandatory minimum green percentage was increased from the previous 10–25% of the plot area. It is specified that the minimum percentage of 25% of the area can be achieved with the mandatory 10% in direct contact and that green areas mean all green areas, including high greenery, green roofs, and greenery above the underground in a certain ratio. In this way, the possibility of using different types of greenery was introduced.

In parallel with that, from 2021, the public company Medijana, which until then only performed the tasks of maintaining greenery and municipal waste, was introduced as the holder of public authorities, i.e., it was given the right to issue conditions for greenery in procedures of implementation of planning documents (creating urban projects and location conditions). In this way, in addition to the planning guidelines, the possibility of implementation was improved.

Changes to the planning documents that followed for the areas of Palilula municipality (Sixth Amendments and Supplements to the GRP of the CM of Palilula—Phase I; Second Amendments and Supplements to the GRP of the CM of Palilula—Phase II), while maintaining a similar principle as for the central area (increase from 10 to 25% of greenery), further specified the technical details and descriptions of green areas, roofs, woody plants, and the like. To stimulate the planting of trees, the precise measure "one apartment—one tree" was introduced, that is, for each newly designed apartment, one sapling should be provided within the plot or in the public area. There is also the definition of "free areas in common use," namely a minimum of 20% of the plot area, of which 15% should be in open space (in contact with the ground, on the ground floor, on open terraces or walkable roofs). These areas do not include green areas, parking spaces, ramps, technical rooms, horizontal and vertical communications, and all areas intended for cars, containers, etc. Free areas in common use should serve the tenants as a whole, for performing various free activities (zones for quiet rest, children's play, spaces for socializing, etc.).

In the area of the municipality of Pantelej (Second Amendments and Supplements to the GRP of the CM of Pantelej—Phase I; First Amendments and Supplements to the GRP of the CM of Pantelej—Phase III West), all measures have been introduced, as in the area of the municipality of Palilula, with additional definition of seedling selection, seedling diameter, and tree spacing, which are defined tabularly through the Greenery Typology. For the undeveloped part of the Pantelej settlement, where a problem (citizen protest) was recorded, due to the complexity and definition of other construction rules as well as subdivision, a zone was designated for further elaboration with a DRP. The tendency to increase the percentage of green areas and to give importance to this is visible in the name itself, as well as in the graphic representation of the zone—mixed-use zone (with larger green areas). The guidelines for further development, defined in this planning document, are given without precise construction rules, which must be determined based on various analyses in the process of developing a DRP. What is given as a clear guideline for further development is precisely the binding minimum percentage of greenery of 30% of the area of the plot/complex, as well as binding uses, among which are public purpose facilities and arranged green areas and sports and recreational areas. Further development of DRPs should be carried out in such a way as to enable an integral approach, harmonization of the ownership structure, and uniform planning of construction and vacant areas. Whether further planning, as well as its implementation, will improve placemaking and the state of greenery remains for further analysis in the future.

Apart from the introduction of individual guidelines through planning documents, some authors point to the need to develop a model that would increase the amount of greenery but would also respect the interests of investors and motivate them to include more green areas in residential buildings (Third Amendments and Supplements to the GRP of the CM of Medijana). This could be achieved by introducing a new parameter in the urban planning legislation that would define greenery in a better and more comprehensive way. This parameter could be defined as a greenness index and would take into account all the greenness on the building plot. The existing minimum percentage of greenery means only green areas on the ground, covered with grass. Trees and high-density greenery such as bushes are completely neglected. In order to improve the quality and quantity of greenery in urban residential areas, and therefore the quality of life of residents, this greenery index should increase the required amount of greenery on construction plots. Models for calculating individual indices should be the subject of further research by landscape architects. They should be very well formulated with simple calculations, and easily applicable in the process of urban and architectural design (Kondić et al. 2018).

Added to that is a similar initiative, which actively advocates the introduction of the ecological index into urban planning (Third Amendments and Supplements to the GRP of the CM of Crveni Krst—Phase I).

The ecological index is a complex urban indicator, whose set value is achieved by a combination of various ecological urban parameters (such as areas of greenery in direct contact with the ground, semi-porous areas, diversity of vegetation, the number and quality of planted trees, the area of a green roof or a green wall), which significantly contributes to environmental comfort with each new construction (Mitić-Radulović 2022).

11.4 Analysis and Results

The paper presents three examples of multi-family residential areas with different characteristics. For easier comparison, the basic data on the analyzed areas are shown in Table 11.3.

The Krive livade settlement was planned and built in the 1970s, during the period of socialism. The construction of the housing complex was in the public sector, on

Table 11.3 Characteristics of the analyzed areas of multi-family housing

	Krive livade	Somborska–Studenička	Vagres City
Appearance of the area (complex) of multi-family housing			
Construction period	1980s	2005–still under construction	2016–2022
Planning	Public sector	Public sector	Private sector
Land ownership	Public	Private	Private
Developer	Public sector (state/social enterprise)	Private sector	Private sector
Share of green areas	73% before transformation 48% today	About 5% at the level of the entire settlement (built-up part of the settlement, not including plots that are still under construction)	About 5.5% at the level of the residential complex

Source Authors; photos: https://free-map.org/

land that was nationalized, that is, state-owned. Green areas made up about 70% of the area intended for housing. In the period of post-socialism, this settlement underwent significant transformations. Green areas were reduced by a third, with the new construction, i.e., today they make up 48% of the area intended for housing. Despite the negative changes, the reduction and fragmentation of green areas, it is learned that inter-block greenery is still to a significant extent preserved and represents an exceptional potential in terms of establishing social interaction among the residents of the settlement, especially today when awareness of the importance of green areas in cities is growing.

In the case of the settlements of Somborska-Studenička and Vagres City, there are significant differences in planning and the way they were created, while the results in space are almost identical.

The settlement of Somborska–Studenička began to develop in the early 2000s with the construction of residential buildings on individual plots, at a time when the area had not yet been thoroughly planned. Urban planning in the initial phase of the transition was still not adapted to the new conditions (private ownership of land, construction on individual plots). The construction of multi-family residential buildings preceded urban planning, which was later adapted to the existing conditions in the area. The shape and size of the blocks are partly the result of adaptation to the illegal construction found. In the spirit of socialist planning, the provisions of the plans concerning green areas were in the form of recommendations, which were not implemented as non-binding. Urban planning parameters were set at the level of a block of buildings bordered by streets and were implemented at the level of a building plot, which gave additional space for deviations. The result of such planning and construction is the absence of public green areas, as well as green areas within housing. In addition, the private ownership of the land and the extremely high degree of development leave no possibility for improving the situation through the development of new green areas.

As for the residential complex Vagres City, which was built about 15 years later, with formal compliance with the planning documentation, the results are identical to those of the Somborska-Studenička settlement. In this case, identically, there are no open public areas except the streets, which is understandable since it is land that is entirely privately owned by the investor. Greenery on the surfaces of other uses, in this case housing, was also reduced to the remains of the space during construction. As in the previous case, the large degree of construction of the area and almost 100% occupancy of free areas by parking do not leave the possibility of increasing the percentage of greenery within the complex. The existing fragmented and scattered green areas have only a limited ecological function and a role in improving the aesthetic characteristics of the rebuilt residential complex, while in terms of rest, recreation, and establishment of social interaction, they have no value nor the possibility of improvement.

The state of green areas within multi-family housing, and the treatment of green areas in the examples shown are the result of multi-layered changes in the transition period, primarily:

- Change of actors in planning and construction—withdrawal of the public sector from housing construction and strengthening of private investors; and at a later stage, the assignment of the preparation of planning documents to private companies, that is, planning in the interest of private investors;
- changing the status of the city's construction land through the amendment of the law and the constitution;
- change in planning method—types of planning documents and level of detail of planning documents;
- privatization and restitution;
- unequal planning and maintenance of undeveloped and built-up areas, preference is given exclusively to built-up space;
- lost connection between planning and implementation;
- lost connection between planning and maintenance.

11.5 Conclusion

Urban green spaces have the potential to be transformational spaces that enrich the social fabric of cities and nurture vibrant, inclusive communities. However, realizing this potential requires a concerted effort in planning, designing, and overcoming obstacles that hinder meaningful social interaction and placemaking. By prioritizing equity, accessibility, and community engagement, cities can harness the power of urban green spaces to create spaces that inspire connection, creativity, and collective well-being. For that, we need better planning documents and guidelines, better implementation as well as continuity in the maintenance of green spaces.

The analysis of three examples of green areas of residential areas from different periods shows that the green areas of settlements that were planned and built in the period of socialism were largely preserved, even after the negative transformations they were exposed to in the post-socialist period. On the other hand, residential construction in the post-socialist period is characterized by the complete neglect of green areas, despite their importance for the quality of the environment, people's health, and the impact on the value of real estate.

The new initiatives shown are encouraging, but they are not systemic solutions, but only individual attempts. What is missing is systematic regulation by law and by-laws, both for green planning in urban areas and for its management.

Also, a systemic change in the way of planning is necessary, integrally, by observing the neighborhood and defining the rules for the neighborhood and not the individual plot, and after that, we can expect results.

It is evident that the divided ownership structure makes it difficult to implement placemaking, but this is not the only reason, because the example of one ownership structure (Vagres City) also shows the absence of the creation of common and green areas within the complex. This proves that there is no systemic solution, as well as penal provisions in case of non-implementation of the planned solutions, which further leads to the conclusion of the absence of monitoring of the performed situation.

The presented examples from the socialist period prove that there is not only potential, but also possibility, but inadequate planning and management indicate that this potential has not been used. Therefore, the main conclusions for such a situation can be stated as follows:

- Absence of legal regulations (laws and other regulations);
- unconnected planning and management processes;
- inadequate hierarchical planning;
- failure to observe space integrally;
- insufficient awareness and expressed need.

References

Action plan for sustainable development of the City of Niš 2015–2020 (2014). https://investnis.rs/wp-content/uploads/2020/04/akcioni-plan-razvoja-grada-nisa-2015-2020.pdf

Backović V (2005) European cities in post-socialist transformation [Evropski gradovi u postsocijalističkoj transformaciji]. Sociology [Sociologija] 47(1):27–44

Bakay E (2012) The role of housing estates' green surfaces in forming the city climate of Budapest. Appl Ecol Environ Res 10(1):1–16

Benkő M (2015) Budapest's large prefab housing estates: urban values of yesterday, today and tomorrow. Hung Stud 29(1–2):21–36

Cilliers EJ, Timmermans W, Van den Goorbergh F, Slijkhuis J (2015) Green place-making in practice: from temporary spaces to permanent places. J Urban des 20(3):349–366

Constitution of SFRY (Official Gazette of SFRY, no. 9/74)

Constitution of the FRY (Official Gazette of FRY, no. 1/92)

Constitution of the Republic of Serbia (Official Gazette of RS no. 1/90)

Crompton JL, Nicholls S (2022) The impact of park views on property values. Leis Sci 44(8):1099–1111

Đekić JP (2022) The influence of changes in the physical and functional structure of multi-storey residential areas on the quality of life of their residents in the conditions of the post-socialist transformation in the example of the city of Niš. Doctoral dissertation, University of Niš. https://nardus.mpn.gov.rs/handle/123456789/21324

Đekić J, Vasilevska L (2021) Characteristics of multifamily housing development in the post-socialist period: case study, the city of Niš. Facta Univ, Ser: Arch Civ Eng 245–260

Đekić J, Dinić Branković M, Mitković P, Igić M, Mitković M (2017) Urban green areas planning and development: an assessment of general urban plans of the city of Niš. Facta Univ, Ser: Arch Civ Eng 15(2):211–224

Đekic J, Đukic A, Vukmirovic M, Đekic P, Brankovic MD (2018a) Thermal comfort of pedestrian spaces and the influence of pavement materials on warming up during summer. Energy Build 159:474–485

Đekić J, Mitković P, Dinić-Branković M, Igić M, Đekić P, Mitković M (2018b) The study of effects of greenery on temperature reduction in urban areas. Therm Sci 22(Suppl. 4):988–1000

Dinic Brankovic M, Igic M, Djekic J, Mitkovic M (2022) Impact of post-socialist vertical extensions of buildings on outdoor microclimate in collective housing areas: a study of Niš, Serbia. Energy Build 265:112081

Dinic-Brankovic M, Djekic J, Mitkovic M, Mitkovic P, Markovic-Brankovic J (2018) Postsocialist restructuring of city subcenters and the role of shopping centers in Nis, Serbia. J Urban Plan Dev 144(2):04018009

DRP of the "Somborska-centar" settlement ("Official Gazette of the City of Niš", no. 106/04)

First amendments and supplements to the general regulation plan of the city municipality of Pantelej—phase III west ("Official Gazette of the City of Niš", no. 63/2023)

General regulation plan of the city of Niš 2010–2025 ("Official Gazette of the City of Niš" no. 43/11), i.e. general regulation plan of the city municipality of Pantelej—phase I ("Official Gazette of the City of Nis" No. 102/2012)

Gradinaru SR, Onose DA, Oliveira E, Slave AR, Popa AM, Gravrilidis AA (2023) Equity in urban greening: evidence from strategic planning in Romania. Landsc Urban Plan 230:104614

Grava S (2007) Urban transport in the Baltic republics. The post-socialist city. Springer, Dordrecht, pp 313–343

Gulin Zrnić V (2017) Green space in socialist and post-socialist Zagreb. In: Green landscapes in the European city, 1750–2010. Taylor & Francis, pp 105–119

Hegedüs J, Mark K, Struyk R, Tosics I (1993) Local options for transforming the public rental sector: empirical results from two cities in Hungary. Cities 10(3):257–271

Hirt S (2008) Landscapes of postmodernity: changes in the built fabric of belgrade and sofia since the end of socialism. Urban Geogr 29(8):785–810

Hirt S (2015) Planning during post-socialism. In: International encyclopedia of the social & behavioral sciences, 2nd edn, vol 18. Elsevier

Hirt S, Kovachev A (2006) The changing spatial structure of post-socialist Sofia. In: Tsenkova S, Budić-Nedović Z (eds) The urban mosaic of post-socialist Europe. Physica-Verlag HD, pp 113–130

Hirt S, Stanilov K (2007) The perils of post-socialist transformation: residential development in Sofia. The post-socialist city: urban form and space transformations in Central and Eastern Europe after socialism. Springer, Netherlands, Dordrecht, pp 215–244

Jayasooriya VM, Ng AWM, Muthukumaran S, Perera BJC (2017) Green infrastructure practices for improvement of urban air quality. Urban For Urban Green 21:34–47

Kondić S, Obradović T, Živković M, Tanić M, Nikolić V, (2018) The treatment of greenery in urban planning documents: residential areas in Niš, Serbia. In: Conference proceedings ICUP 2018, pp 365–372

Kronenberg J, Haase A, Łaszkiewicz E, Antal A, Baravikova A, Biernacka M et al (2020) Environmental justice in the context of urban green space availability, accessibility, and attractiveness in postsocialist cities. Cities 106:102862

Krstić V, Medvedev M (1983) Urban development of Niš (1945–1980): planning experiences, state and perspectives [Urbani razvoj Niša (1945–1980): Planerska iskustva, stanje i perspektive]. Niš Anthology [Niški zbornik] 12:64–82

Law for Urban Greenery (Official Gazette of the Republic of Macedonia, No. 11/2018)

Law on Construction Land (Official Gazette of RS, No. 44/1995)

Law on Construction Land (Official Gazette of the SR of Serbia, No. 23/1990)

Law on Nationalization of Rental Buildings and Construction Land (Official Gazette of FNRY, No. 52/58)

Law on Planning and Construction (Official Gazette of RS, No. 47/2003)

Law on Planning and Construction (Official Gazette of RS, No. 72/2009, 81/2009, 64/2010 (Decision of Constitutional Court), 24/2011, 121/2012, 42/2013 (Decision of Constitutional Court), 50/2013 (Decision of Constitutional Court), 54/2013 (Decision of Constitutional Court), 98/2013 (Decision of Constitutional Court), 132/2014, 145/2014, 83/2018, 31/2019, 37/2019 (other law), 9/2020, 52/2021)

Law on the Establishment and Maintenance of Green Areas (Official Gazette of the SR of Serbia, No. 8/72)

Lottrup L, Grahn P, Stigsdotter UK (2013) Workplace greenery and perceived level of stress: benefits of access to a green outdoor environment at the workplace. Landsc Urban Plan 110:5–11

Mandič S (2010) The changing role of housing assets in post-socialist countries. J Housing Built Environ 25:213–226
Marcuse P (1996) Privatization and its discontents: property rights in land and housing in the transition in Eastern Europe. In: Cities after socialism: urban and regional change and conflict in post-socialist societies, pp 119–191
Marin V, Berescu C, Macri Z (2023) An uncertain future: prospects for Bucharest's large housing Estates. J Housing Built Environ 38(1):101–119
Medvedev M (2018) Urban tradition of Niš in plan views [Urbana tradicija Niša u planskim prikazima]. Society of Architects of Niš [Društvo arhitekata Niša], Niš
Melichar J, Kaprová K (2013) Revealing preferences of Prague's homebuyers toward greenery amenities: the empirical evidence of distance–size effect. Landsc Urban Plan 109(1):56–66
Mitić-Radulović A (2022) Possibilities of applying the ecological index in the planning of Belgrade [Mogućnosti primene ekološkog indeksa u planiranju Beograda]. Center for Experiments and Urban Studies—CEUS [Centar za eksperimente i urbane studije]
Moradpour M, Hosseini V (2020) An investigation into the effects of green space on air quality of an urban area using CFD modeling. Urban Clim 34:100686
Navarrete-Hernandez P, Laffan K (2019) A greener urban environment: designing green infrastructure interventions to promote citizens' subjective wellbeing. Landsc Urban Plan 191:103618
Nesbitt L, Meitner MJ, Sheppard SR, Girling C (2018) The dimensions of urban green equity: a framework for analysis. Urban For Urban Green 34:240–248
Obradović T, Mitković P (2012) The development of urban legislation in Serbia and England. Facta Univ, Ser Arch Civ Eng 10(3):315–326
Obradović T, Polić D (2024) Analysis of regulations and practice of urban planning. Recommendations for the improvement of the regulatory framework and methodology for the preparation of planning documents at the local level with the aim of achieving a better quality of the built environment [Analiza propisa i prakse izrade urbanističkih planova. Preporuke za unapređenje regulatornog okvira i metodologije za izradu planskih dokumenata na lokalnom nivou s ciljem dostizanja boljeg kvaliteta građene sredine]. In: Standing conference of towns and municipalities [Stalna konferencija gradova i opština], Belgrade
Price A, Jones EC, Jefferson F (2015) Vertical greenery systems as a strategy in urban heat island mitigation. Water Air Soil Pollut 226:1–11
Project for Public Spaces (PPS) (2007) What is placemaking? www.pps.org/article/what-is-placemaking
Rađenović T, Radivojević V, Krstić B, Stanišić T, Živković S (2022) The efficiency of health systems in response to the COVID-19 pandemic: evidence from the EU countries. Problemy Ekorozwoju—Probl Sustain Dev 17(1):7–15. https://doi.org/10.35784/pe.2022.1.01
Roy S, Byrne J, Pickering C (2012) A systematic quantitative review of urban tree benefits, costs, and assessment methods across cities in different climatic zones. Urban For Urban Green 11(4):351–363
Second amendments and supplements to the general regulation plan of the city municipality of Palilula—phase II ("Official Gazette of the City of Niš", No. 63/2023)
Second amendments and supplements to the general regulation plan of the city municipality of Pantelej—phase I ("Official Gazette of the City of Niš", no. 139/2022)
Sixth amendments and supplements to the general regulation plan of the city municipality of Palilula—phase I ("Official Gazette of the City of Niš", No. 26/2023)
Study on the development of the local economy of the city of Niš (2005). https://www.gu.ni.rs/wp-content/uploads/file/studija-razvoja-lokalne-ekonomije.pdf
Sýkora L (1994) Local urban restructuring as a mirror of globalisation processes: Prague in the 1990s. Urban Stud 31(7):1149–1166
Szczepańska A, Krzywnicka I, Lemański G (2016) Urban greenery as a component of real estate value. R Estate Manag Valuat 24(4):79–87

Szelényi I (1996) Cities under socialism—and after. In: Andrusz G, Harloe M, Szelényi I (eds) Cities after socialism: urban and regional change and conflict in postsocialist societies. Blackwell Publishers, Oxford, pp 286–317

The World Bank (2022) Urban population (% of total population). https://data.worldbank.org/indicator/SP.URB.TOTL.IN.ZS?locations=XU

Third amendments and supplements to the general regulation plan of the city municipality of Medijana ("Official Gazette of the City of Niš", No. 129/2021)

Third amendments and supplements to the general regulation plan of the city municipality of Crveni Krst—phase I ("Official Gazette of the City of Niš", No. 85/2022)

Treija S, Bratuškins U (2019) Socialist ideals and physical reality: large housing estates in Riga, Latvia. Housing estates in the Baltic countries. Springer, Cham, pp 161–180

Tsenkova S (2006) Beyond transitions: understanding urban change in post-socialist cities. In: The urban mosaic of post-socialist Europe: space, institutions and policy. Physica - Verlag. A Springer Company, pp 21–50

Tsenkova S, Polanska DV (2014) Between state and market: housing policy and housing transformation in post-socialist cities. GeoJournal 79:401–405

Turvey RA (2019) Place-making and sustainable community development. In: Turvey RA, Kurissery S (eds) Intellectual, scientific, and educational influences on sustainability research. IGI Global, pp 253–272

Tuvikene T (2019) Between community and private ownership in centrally planned residential space: governing parking in socialist housing estates. In: Housing estates in the Baltic countries: the legacy of central planning in Estonia, Latvia and Lithuania, pp 321–336

United Nations, Department of Economic and Social Affairs (UNDESA) (2018) 68% of the world population projected to live in urban areas by 2050, says UN. https://www.un.org/uk/desa/68-world-population-projected-live-urban-areas-2050-says-un.

United States Census Bureau (2022) Nation's urban and rural populations shift following 2020 census. https://www.census.gov/newsroom/press-releases/2022/urban-rural-populations.html

Uvalić M (2010) Serbia on the eve of transition. Serbia's transition: towards a better future. Palgrave Macmillan, New York, pp 14–44

Vagres City. https://vagres.rs/en/vagres-city/

Vasilevska L, Vranić P, Marinković A (2014) The effects of changes to the post-socialist urban planning framework on public open spaces in multi-story housing areas: a view from Nis, Serbia. Cities 36:83–92

Vasilevska L, Milanovic D, Nikolic M, Vranic P, Milojkovic A (2015) "Garage capitalism" as a form and process of post-socialist urban changes: its pace, intensity and structural characteristics. A case study of Nis, Serbia. Habitat Int 48:149–158

Vasović D, Rađenović T, Živković S (2023) Public utility systems in the Republic of Serbia: current state and future perspectives with particular regard to economic issues and SDG 6 targets. In: Gjorchev J, Malcheski S, Rađenović T, Vasović D, Živković S (eds) Transformation and efficiency enhancement of public utilities systems: multidimensional aspects and perspectives. Business Science Reference, Hershey, PA, pp 1–26. https://doi.org/10.4018/978-1-6684-7730-4.ch001

Vranjanac Ž, Rađenović Ž, Rađenović T, Živković S (2023) Modeling circular economy innovation and performance indicators in European Union countries. Environ Sci Pollut Res 30(34):81573–81584. https://doi.org/10.1007/s11356-023-26431-5

Zellner M, Massey D, Minor E, Gonzalez-Meler M (2016) Exploring the effects of green infrastructure placement on neighborhood-level flooding via spatially explicit simulations. Comput Environ Urban Syst 59:116–128

Živković S, Milutinović S, Rađenović T (2022). The impact of COVID-19 on the environment. In: 19th international conference "Man and Working Environment" occupational and environmental safety engineering and management (OESEM), November, 24–25, 2022. Faculty of Occupational Safety, University of Niš, pp 219–224

Data-Policy Interaction

Placemaking of urban green areas involves the intentional design and management of public spaces to enhance the quality of life and well-being of residents while fostering community engagement and sustainability. It intersects with EU policies, especially with the European Green Deal and Urban Agenda for the EU, as well as with the UN SDGs, particularly SDG 11: Sustainable Cities and Communities, SDG 13: Climate Action, and SDG 15: Life on Land. The European Green Deal, the EU's flagship initiative to make the EU's economy sustainable and climate-neutral by 2050, is well aligned in its goals with the placemaking of urban green areas by promoting biodiversity, reducing greenhouse gas emissions, and enhancing the resilience of urban environments to climate change. Besides, placemaking of urban green areas contributes to several priority areas of the Urban Agenda, including green infrastructure, circular economy, and climate adaptation. When it comes to the UN SDGs, urban green areas directly contribute to SDG 11 by making cities and human settlements inclusive, safe, resilient, and sustainable. These green spaces enhance the livability of cities, promote social cohesion, and improve public health and well-being. Besides, they play a crucial role in climate change mitigation and adaptation. Trees and vegetation in urban parks and green spaces help sequester carbon dioxide, mitigate urban heat islands, and reduce air pollution, thereby contributing to SDG 13 targets. Furthermore, placemaking efforts that prioritize the conservation and restoration of urban green spaces contribute to SDG 15 by protecting terrestrial ecosystems, biodiversity, and natural habitats within urban environments. Therefore, by integrating placemaking principles into urban planning and policy frameworks, governments, local authorities, and stakeholders can create more resilient, inclusive, and sustainable cities for current and future generations.

Tanja Obradović is a Ph.D., in Architecture, specializing in urban planning. Her research focuses on building regulations, changes in space, with a focus on applying complexity theory to understand and shape urban systems. Since 2005, she has been working at the Ministry of Construction, Transport, and Infrastructure in the urban planning inspection department, with a break from 2021 to 2023 when she served as the chief city planner of Niš. She participated in the World Urban Forum in Katowice, presenting public spaces in Niš. She is the founder and active organizer of the international scientific conference ICUP, which takes place every two years in Niš. She is a co-author of the document "Analysis of regulations and practices in the development of urban plans" SCTM (2023). She worked as a UN-Habitat coordinator for the localization of sustainable development goals in preparing the Voluntary Local Review of the City of Niš (2023–2024). She also coordinated the pilot of a new quality of life index for the City of Niš within the UN-Habitat Quality of Life initiative (2024).

Jelena Đekić is an assistant professor at the University of Niš, Faculty of Civil Engineering and Architecture, with research experience in the field of urban and spatial planning. Her research focuses on the post-socialist transformations of urban fabric, urban quality of life, public open spaces, outdoor thermal comfort, and legislation in urban and spatial planning. She is the author and co-author of more than 50 papers published in national and international journals and presented at scientific conferences. As a planner at the Institute of Urban Planning City of Niš, she participated in the preparation of several spatial and urban plans and strategic environmental impact assessments. She participated in two projects of the German Academic Exchange Service (DAAD) as a project assistant. She is a researcher in the ongoing scientific research project Reclaiming Public Open Space in Residential Areas: Shifting Planning Paradigms and Design Perspectives for a Resilient Urban Future, funded by the Science Fund of the Republic of Serbia within the Prisma Program (2024–2027).

Tamara Rađenović is an associate professor at the Faculty of Occupational Safety, University of Niš, Serbia. She earned her MA in Finance and Investments with Merit from Nottingham University Business School, UK, in 2007, and Ph.D. in Economics from the Faculty of Economics, University of Niš, Serbia, in 2018. She teaches safety economics in undergraduate studies, and economics of environmental protection, in master studies. Her areas of interest include intellectual capital management, performance management, economic development, financial markets, public–private partnerships, safety economics, environmental economics, urban governance, sustainability, and real options. She has published over 100 papers in the relevant journals indexed in WoS and Scopus including Environmental Science and Pollution Research, Renewable and Sustainable Energy Reviews, Sustainability, Journal of Balkan and Near Eastern Studies, Acta Oeconomica, Problems of Sustainable Development, as well as, in national journals. Also, she is a co-author of two national monographs in the field of Strategic and Operational Intellectual Capital Management and co-editor of the two international monographs in the field of Occupational Safety Management and Public Utilities Systems Efficiency.

Milena Dinić Branković, Ph.D., is an associate professor at the University of Niš, Faculty of Civil Engineering and Architecture. In the area of urban planning and design, her research focuses on residential land use, central zones of urban/rural settlements, public open spaces, outdoor thermal comfort, and post-socialist transformations of urban fabric. She has published more than 80 papers, three chapters in international monographs, and two books. She is the principal investigator of the scientific research project *Reclaiming Public Open Space in Residential Areas: Shifting Planning Paradigms and Design Perspectives for a Resilient Urban Future*, funded by the Science Fund of the Republic of Serbia within the Prisma Program (2024–2027). She is a section editor of the journal *Facta Universitatis, Series Architecture and Civil Engineering*. She works as an UN-Habitat consultant for the localization of sustainable development goals on the preparation of the Voluntary Local Review of the City of Niš (2023–2024). She is an alumna of the *Junior Faculty Development Program*, an international educational faculty exchange program funded by the Bureau of Educational and Cultural Affairs of the United States Department of State (2010).

Milica Igić, Ph.D., is a teaching assistant at the Chair for Urban and Spatial Planning at the Faculty of Civil Engineering and Architecture, University of Niš. She has research experience in the field of urban and spatial planning, and her main research focuses on built areas and climate change, the revitalization and development of rural areas, and the implementation of nature-based solutions in urban and rural open spaces. She is the author and co-author of a large number of scientific papers; she also published one and was co-author in two chapters in International Monograph. She is a member of the organizing committee of a few International Conferences and International Exhibition organized by the Faculty of Civil Engineering and Architecture. She participated in the DAAD International scientific project in 2020 and is a researcher in the ongoing scientific research project *Reclaiming Public Open Space in Residential Areas: Shifting Planning Paradigms and Design Perspectives for a Resilient Urban Future*, funded by the Science Fund of the Republic of Serbia within the Prisma Program.

Chapter 12
Enhancing Natural and Cultural Heritage Through Sustainable Tourism and Improved Accessibility to Support Economic Development of Sensitive Regions in the Balkan and Eastern European Countries

Ilenia Pierantoni, Nora Annesi, and Massimo Sargolini

Abstract The chapter presents the main results of three research projects developed in the last years in the context of Balkan and Eastern European Countries: Interreg South East Europe Access2Mountains, which regards Alpine and Carpathian regions; Interreg Italy–Croatia MADE IN-LAND and Interreg Italy–Croatia BOOST5, which interests Italian and Croatian Regions. These projects' main objective is to enhance the protection and valorization of natural and cultural unexploited capital of inland sensitive regions by means of environmentally friendly tourism development and improved (sustainable) accessibility. The chapter will provide insights into the regional development of regions understudy, focusing on specific local challenges: on the one hand, the fragility of the tourism sector being highly fragmented, imbalanced, and still with low digital intensity; on the other, the lack of sustainable alternative mobility solutions to serve tourist itineraries. Project results demonstrate the importance of data-driven policymaking and the role of data in understanding specific criticalities and in shaping and guiding decision-making. In this framework, the projects foresee the deployment of a systemic approach and common methodologies for natural and cultural enhancement to: eliminate main obstacles limiting the full development of sustainable tourism offers; promote community-driven ideas and new market opportunities through capacity-building plans about innovative approaches, sustainability and managerial skills; promote synergies among diverse stakeholders and territories to speed up successful implementation of sustainable accessibility policies and environmental awareness; pave the way for the next strategic steps such as

I. Pierantoni (✉) · M. Sargolini
School of Architecture and Design, University of Camerino, Camerino, Italy
e-mail: ilenia.pierantoni@unicam.it

N. Annesi
Department of Business, Law, Economics and Consumer Behaviour, IULM University, Milan, Italy

J. J. Stanković et al. (eds.), *Regional and Urban Challenges and Opportunities for Data-Policy Interactions*, The Urban Book Series,
https://doi.org/10.1007/978-3-032-08592-4_12

green accreditation or zero-impact tourism labels and building on next programming period opportunities.

Keywords Cultural heritage · Sustainable tourism · Sustainable mobility · Cross—border cooperation · Place-based policies · SDGs

12.1 Introduction

The presented chapter delves into the multifaceted realm of sustainable tourism development and enhanced accessibility, focusing on the sensitive dynamics of inland regions within the Balkans and Eastern European countries. The chapter is based on the shared commitment to protecting and valorizing the under-valorized natural and cultural capital within these inland regions.

These topics are identified as critical aspects on which the institutions, the academics, and all the other local actors should focus their efforts. Indeed, the European Cohesion strategy has highlighted the importance of having equal socio-economic development among cities and inner areas (EU 2007; Battaglia et al. 2019). In the same vein, the 2030 Agenda has stressed the importance of implementing actions to protect the planet through a sustainable consumption and production system, preserve natural resources, and live prosperously in harmony with nature. In particular, with SDG 2, SDG 11, SDG 8, and SDG 15, the 2030 Agenda explains priorities related to the importance of ensuring "*ensure sustainable food production systems and implement resilient agricultural practices that increase productivity*" (Target 2.4) to "*double the agricultural productivity and incomes of small-scale food producers*" (Target 2.3)" also by supporting "*a positive economic, social and environmental links between urban, peri-urban and rural areas by strengthening national and regional development planning*" (target 11.a). Moreover, the 2030 Agenda add the importance of implementing "*policies to promote sustainable tourism that creates jobs and promotes local culture and products*" (Target 8.9) while ensuring "the conservation, restoration and sustainable use of terrestrial and inland freshwater ecosystems and their services" (Target 15.1).

All the issues above are part of this chapter. Indeed, the presented results and the critical reflections have been made on the basis of 3 Interreg Research Projects. Interreg is a European Union Founding Programme to support cooperation across borders through projects and joint initiatives. Our exploration focuses on the results of three Interreg projects: Interreg South East Europe Access2Mountains, the Interreg Italy–Croatia MADE IN-LAND, and the Interreg Italy–Croatia BOOST5.

The primary objective of these projects is to champion environmentally friendly tourism development while concurrently improving sustainable accessibility. This dual approach not only seeks to unlock the economic potential of these sensitive regions, but also endeavors to safeguard and valorize their rich natural and cultural heritage. Through this approach, we discuss the delicate balance between economic prosperity and the responsible stewardship of these precious resources.

The chapter explores the dynamics occurring in different contexts of Italian, Balkans, and Eastern regions, overcoming borders to create a cohesive framework for sustainable development and making results replicable in similar contexts. In line with the Interreg framework, these projects enclose the spirit of collaboration, bringing together diverse stakeholders, including governmental bodies, academic institutions, local communities, and businesses. Through the presentation of empirical cases, the chapter aims to transcend the conventional boundaries of regional development, fostering a holistic and inclusive approach that respects the unique attributes of each context while promoting shared prosperity.

The following chapter will explain the peculiarities of three research projects, exploring their specific objectives, methodologies, and, most importantly, the tangible outcomes that contribute to the sustainable development of the inland regions of Balkan and Eastern Europe. The chapter dissects the commonalities that bind these projects and sheds light on their transformative potential for these sensitive regions' cultural and natural heritage.

12.2 Research Projects and Major Challenges

12.2.1 ACCESS2MOUNTAIN

The first project presented here is ***ACCESS2MOUNTAIN—Sustainable Mobility and Tourism in Sensitive Areas of the Alps and the Carpathians*** (hereafter Access2Mountain).[1] The project is an Interreg South East Europe developed in the framework of the Alpine and Carpathian Conventions, which is a transnational convention aimed at pursuing comprehensive policy and cooperation to promote environmental protection and sustainable development of mountain regions, addressing environmental challenges and improving the quality of life for current and future generations. In May 2011, the ACCESS2MOUNTAIN project started as a transnational collaboration between 32 partners and observers from 17 countries in the Alps and the Carpathians. Funded by the Southeast Europe Programme, Access2Mountain was developed to promote the improvement of accessibility in mountain areas as a crucial factor in their sustainable social, economic, and cultural development, with particular regard for tourism. In the meanwhile, the project aimed at bringing to light methods, models, and tools for the containment and reduction of the negative impacts of traffic, infrastructures, and tourist flows on the environment, landscape,

[1] "Access2Mountain" - Sustainable Mobility and Tourism in Sensitive Areas of the Alps and the Carpathians (SEE/B/0007/3.1/X).

Project partnership: Environment Agency Austria (LP), National Park Gesäuse, Mostviertel-Tourism Ltd., Regional Development Agency of North Hungary, University of Camerino, EURAC research, CJIT Maramures, ARR-KE, RARR-PL, Alpine Convention, MATTM, Carpathian Foundation Ukraine, MoT-SI, NÖVOG, RPK-Zajecar, CAA, OSD, TBAR, PKS, Fedecrail, RDA Svidnik. Scientific responsible for the Unicam Research Unit: Prof. Massimo Sargolini.

www.access2mountain.eu.

and natural/cultural heritage, particularly in the most sensitive areas of Southeastern Europe. In this framework, the project's specific objectives concerned *(i)* the development of knowledge, awareness, and management capacity among local actors for the development of soft transport policies on different levels and for potential users; *(ii)* the definition of feasibility models aimed at increasing soft tourist mobility in sensitive pilot mountain regions; *(iii)* the implementation of a project of innovative, attractive, and competitive small railways and intermodal transport systems (in the Alpine and Carpathian areas with similar problems but different backgrounds) for tourism development, everyday mobility and in some areas also for environmentally sustainable freight transport; *(iv)* support the reduction of greenhouse gases, pollutant emissions and segmentation related to transport, in view of improving the quality of the environment.

To pursue the above objectives required multi-level and integrated policies based on transnational cooperation, particularly for strengthening the offers, facilities, and infrastructures for tourism, on the one hand, and, on the other hand, raising the benefits and opportunities deriving from tourism. The project served as a prime example of the benefits of transnational cooperation. It involved exchanging best practices, providing practical know-how, and forming long-term partnerships. It aimed at improving cross-border and transnational multimodal transport systems, emphasizing cycling as a sustainable travel mode. Major lessons learned from the project were:

- Transnational cooperation is key: sustainable solutions in tourism and transport often require collaboration across borders. National initiatives alone may not be sufficient; broader, transnational programmes can address the complexities of these sectors more effectively;
- Learning from best practices: sharing experiences and best practices at the transnational level can provide valuable insights. The project showed that regions can significantly benefit from such exchanges, learning from each other's successes and challenges;
- Importance of multimodal transport systems: developing a cross-border and transnational multimodal transport system, including cycling, is vital. Such systems can significantly enhance sustainable mobility, offering environmentally friendly alternatives and improving accessibility to remote regions;
- Innovative mobility solutions: the implementation of innovative and inclusive mobility solutions, such as demand-responsive transport and car-sharing, can address the unique needs of diverse populations in different regions, particularly in remote or mountainous areas;
- Integrating tourism with sustainable transport: viewing cycling and other sustainable transport modes as integral parts of the tourism experience, rather than just recreational activities, can transform tourist regions and offer more comprehensive experiences;
- Regional development through structural funds: utilizing funds like the ERDF to support regional development, especially in sensitive areas like mountain regions,

is crucial. These funds can help overcome geographical and infrastructural challenges, thus improving living conditions and economic prospects;
- Economic and social benefits: sustainable mobility plans contribute to environmental conservation and may have positive economic and social impacts. They can promote regional businesses, create job opportunities, and enhance the overall quality of life of residents;
- Effective project management and communication: the success of such transnational projects hinges on effective leadership and communication. Clear goals, consistent strategies, and open communication channels are essential for the smooth execution of these projects.

These lessons underscore the complexity and interconnectedness of tourism, transport, and sustainability, highlighting the necessity of collaborative, innovative, and well-managed approaches to address these global challenges.

In this way, the Decision Support System (DSS) for the Regional Mobility Management developed in this research can be a scientific tool to support regional practitioners and policy makers during the decision-making process for the sustainable development of their regional transport system.

The DSS has been defined by encouraging the possibility of stimulating the reflection of decision makers, producing alternative scenarios and trying to evaluate them. In this perspective, the tool is an additional support for decision makers that are strengthen in the political responsibility of their decisions.

Particular attention was paid to the identification of indicators and parameters to be used in the evaluation of the current state and then of the project proposal.

12.2.2 Made In-Land[2]

Following up the Access2Mountain Research Project, our research team developed another Interreg Project, ***MADE IN-LAND—MAnagement and DEvelopment of INLANDs*** (hereafter Made in-Land), focusing on Italian and Balkan regions in the framework of the Italy–Croatia Programme and focusing on two specific challenges: *(i)* the fragility of the tourism sector being highly fragmented, imbalanced, and still with low digital intensity; *(ii)* the lack of sustainable alternative mobility solutions to serve tourist itineraries and diffuse natural and cultural heritage.

As for the Access2Mountain Project, Made in-Land focuses on the disbalances that characterize the territory. The Italy–Croatia Programme area, in Italy and Croatia,

---

[2] "MADE IN-LAND"—MAnagement and DEvelopment of INLANDs.

Project partnership: Marche Region (LP); University of Camerino; Superintendence Archeology Fine Arts and Landscape of Marche Region; Municipality of San Leo; Municipality of Riccia; Molise Towards 2000; University of Zadar; Zadar County Development Agency Zadra Nova; Authority for the Management of Parks and Biodiversity—Romagna; Azrri—Agency for Rural Development of Istria Ltd. Pazin.

Scientific responsible for the Unicam Research Unit: Prof. Massimo Sargolini.

https://programming14-20.italy-croatia.eu/web/madeinland.

is composed of two very different but interrelated parts of territory: the coastal belt and the hinterlands. In the first, there is a majority of cities and great urban systems, whereas in the second, there is a prevalence of small towns, villages, and rural settlements, often lacking some basic services. Most of these smaller centers are in fact far away from the main cities, and each inland area is unique and very dissimilar from the others: *specialized*, in a sense, depending on the particular setting and the specific process of local development—depending, in other words, on the available resources and the correlated human activities. Regarding the relationship between the coast and the hinterland, and with regard to their own specific "state of health," similar and comparable dynamics, if not symmetrical, have occurred and are presently occurring in the two countries. On the coastline, the seaside resorts and the variety of natural ecosystems—quite rich yet in biodiversity, beautiful sceneries, and habitats—still provide crucial services and goods to local communities, tourists, and a wider part of the population. Besides being a key factor for the quality of life, the natural environment is indeed one of the primary drivers of economic development and probably the most important element of the territorial brand. However, the attractiveness of coastal areas has led and is leading, in the end, to counterproductive phenomena, such as mass tourism, land take, and urbanization, all due to human solid pressure.

On the other hand, inland areas have witnessed for over thirty years the opposite: a process of gradual marginalization consisting of demographic decrease and aging of the local population; abandonment and decay of local resources; negative economic trends; and reduction of both public and private services (e.g., health care, education, mobility) (Barca 2009; Pierantoni et al. 2020). As in a vicious circle, if the population does fall, the pro-capita costs for ensuring the collective services will increase; therefore, there will soon occur a lack of some needed services and, in turn, the quality of life will decline, finally discouraging potential newcomers from resettling in the hinterlands. Besides, the polarization of the economy to the advantage of the largest urban areas—meaning the shifting of the most profitable activities toward the coastline—implies that the average taxable income in the hinterlands is much lower than on the coast. In addition to this, managing authorities showed limited capacities in conceiving innovative solutions and profound weaknesses in strategic regional planning (sometimes even incapable of cooperating or establishing fruitful, durable partnerships between each other). However, despite all difficulties, inland areas hold a good deal of endogenous resources, either abandoned and forgotten or still unexploited, because they are not even recognized or properly valued (Sargolini et al. 2014; Barca 2014; Canale et al. 2019). Some examples are very precious know-how about traditional arts and crafts, natural environments of high ecological and landscape value, and strong potential in the energy and primary sectors. In this framework, Made in-Land focused on the natural and cultural heritage of inland areas, rich in endogenous resources abandoned, forgotten, or still unexploited and, therefore, with a high development potential for the whole programme area. The project identified the most important natural assets characterizing the areas, the habitats, and the rich biodiversity, already recognized and included in the Natura 2000 network, together with the ecosystem services provided, which support an extraordinarily diversified and valuable agri-food system (Dastgerdi et al. 2020). At the same time, local ecosystems offer an ample supply of renewable energy resources and (green) job opportunities, which need to be exploited through environment-friendly solutions and technologies (Barca 2014; Battaglia and Annesi 2018; Cetara and Pierantoni 2014). As for the cultural assets, the project included both the "tangible" cultural heritage (artworks, monuments, historical places, industrial installations, artifacts, etc.) and the intangible cultural resources (social values, traditions, customs and practices, aesthetic and spiritual beliefs, artistic expression, language, and other aspects of human activity), in particular, the know-how which is at the base of the local typical productions (included the agri-food supply chains) and handicrafts. In solid synergies with the Italian National Strategy for Inland Areas (S.N.A.I.), which sets the background to efficiently promote inland economic and social growth and enhance sustainable development, the main project objective was to develop a cross-border strategy based on win–win cooperation between coastal areas and hinterlands to decrease human pressure on the firsts, while unleashing the potential of natural/cultural assets of the seconds. The strategy tried to go beyond the existing national approaches to inland areas, which often pay too much attention to inland conditions and lose the potential of working on synergies with other areas. The project strategy was to focus on sustainable tourism as a means for local development: the idea was that tourism not only creates growth in the tertiary sector but also has a positive multiplier effect in the primary and secondary sectors (Cetara and Pierantoni 2014; Leask 2016; Sargolini et al. 2020; De Siano and Canale 2022). At the same time, cultural and natural tourism is strictly dependent on the specific attractiveness of the territory, and, indeed, it fosters the conservation of local heritage together with its valorization (Sargolini and Gambino 2014; Canale et al. 2019). This result could be achieved only by connecting activities, which are based upon the specific hinterland 's natural and cultural resources resulting from a long-lasting diversification and specialization process: the project applied a participatory approach thanks to the establishment of local stakeholders committees aimed at strengthening the cooperation and setting up new forms of organization between all the actors of local value chains. Moreover, the committees were based on a quadruple helix approach to support innovation processes: they included not only the local economic actors but also public institutions, research bodies, CSOs and local communities, and other technical bodies in the field. The project strategy for the sustainable economic and territorial development of inland areas was developed with the active partnership of such committees which, at the end of the project, were founded as permanent observatories for the application and evaluation of the project's Action Plan.

The project demonstrated that inland areas need to set up durable partnerships between the local communities and related managing authorities to cooperate and scale up their development plans while coordinating/improving the provision of services intended for local communities. It was evident that inland areas not only need a more robust cooperation of local actors toward developing a strategy for valorizing natural and cultural heritage through sustainable tourism, but they also need to appeal to specific market demands. They need to be reconnected to significant trade and production flows to develop job opportunities and enhance population well-being, which are preconditions for the conservation of natural and cultural assets in the long term. Considering their potential as mass-tourism destinations, the strategy underpinned by the project was based on *win–win* cooperation with coastal areas: fostering cooperation between inland and coastal areas, in particular through sustainable tourism activities, will reduce, on the one hand, the human pressure on the coast, and will unleash, on the other, the potential of inland natural assets and cultural heritage (in terms of economic development) connecting them with new and wider markets.

The strategy developed by the project and adopted by the regional authorities involved in the project goes beyond the existing national strategies for inland areas, which are focused on internal conditions, weaknesses, and opportunities while poorly analyzing the potential of working on synergies with other areas. Made in-Land developed new solutions focusing on the potential natural, cultural, productive, and functional interdependencies between the hinterland and coastal areas: environmental and cultural continuity between the two areas has been poorly valorized in the past; at the same time, they represent a key resource in terms of touristic accessibility (physical and virtual) and integrated destination management.

Overall, the project implemented five pilot actions, three of them located in Italy, which were identified as gateways to inland areas:

1. The Project "O.S.I.C.—One Stop Information Center": Muccia Municipality (Marche area), located in the highest part of Chienti Valley, in the central part of Italy. The area boasts a rich historical and cultural heritage and is characterized by its proximity to urban centers, with cities like Macerata and Ancona playing significant roles in shaping the socio-economic dynamics of the region. Through the project, the vision of valorization of inlands areas' resources was developed by implementing an innovative tourism information center based on the digitalization of local services. The project focused on four objectives: valorization of the cultural and natural heritage; encouragement to enjoy the territory in a "slow" way; integration of local services (heritage, food and wine, hospitality); the development of ICT technologies for communication and information. The OSIC is hosted in the Mulino da Variano, an ancient mill at the entrance of Muccia, that is strategically located on the route connecting the sea to the inland area. OSIC Muccia represents a place of exchange and integration, where fast meets slow, and physical meets digital to reduce the distance (both physical and digital) between the coastal area and the inland (Fig. 12.1).

2. The Virtual Ecomuseum Project for the Municipality of San Leo—Ente Parchi Emilia Romagna. San Leo, situated in the picturesque Emilia Romagna region, shares characteristics with Fortone in terms of its relative distance from urban areas. The seclusion of San Leo contributes to a distinct socio-economic trend and tourism vitality. The region's historical significance and natural beauty, combined with limited accessibility, shape the dynamics of sustainable tourism and economic development. Ecomuseum of San Leo represents a privileged gateway to the historical, naturalistic, and environmental riches of the surrounding area. Travelers who decide to make a stop here will have the privilege of meeting

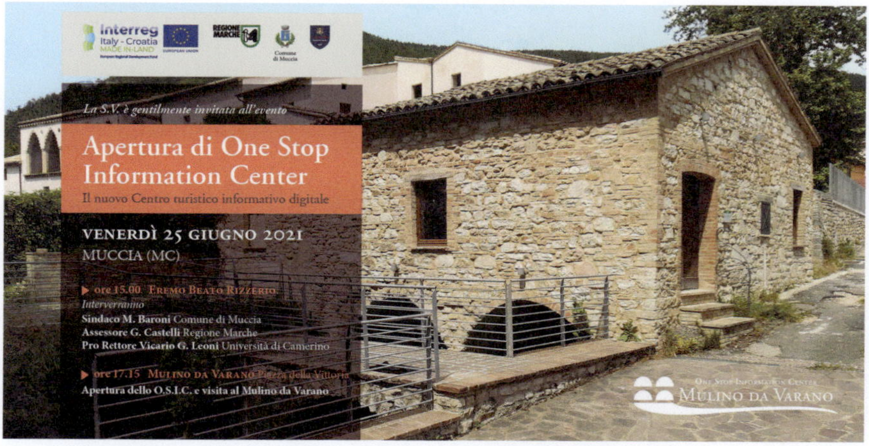

Fig. 12.1 O.S.I.C.—Opening of the one stop information center": Muccia municipality (Marche area), 25 June 2021

three illustrious witnesses, also pilgrims to San Leo: San Francesco, Dante, and Cagliostro. Through their stories, the visitors can discover how all the cultural and architectural heritage of San Leo is born from centuries and centuries of history and from events not infrequently shrouded in legend, if not mystery. The journey begins within the walls of the Fortress, at the top of the fortress. From here, the visitor can look at the ridges of the Tausani Mountains, and imagine the richness of life and history that, from Roman times to today, has animated these places that unfold between rolling hills and steep cliffs. It is a unique territory because it is on the border between two different biogeographical regions, unique for its inaccessibility and for its architectural jewels. Through an immersive, engaging, and exciting experience divided into several rooms, the visitor becomes aware of the identity of the land he is crossing and can continue his journey with a renewed soul to discover the beauty and typicality surrounding Musleo.

3. The Virtual Ecomuseum Project for the Municipality of Riccia and GAL Molise. Fortone, nestled in the Molise region, epitomizes a more secluded setting compared to other pilot areas. The region's low accessibility and distance from major urban centers contribute to its unique socio-economic trends. The challenge of limited connectivity may impact tourism vitality, presenting opportunities for sustainable development that align with the preservation of natural and cultural heritage. Ecomuseum of Riccia Terr@ di Passo virtual Ecomuseum is a large project aimed at promoting and enhancing the cultural and natural heritage of the Riccia area and the municipalities of the Fortore River Valley. 18 municipalities have been involved in the Ecomuseum development. The "Terr@" is the virtual place where this territory is told through the story of an ancient noble family, the de Capua, that this land has controlled, defined, and modified. The Pass is that of shepherds, merchants, and people who have crossed these places.

Fig. 12.2 Terr@ di Passo virtual Ecomuseum

The passage has built the landscape; the road network gave birth to communities. The communities are the heart of this story: from listening to them comes the proposal of knowledge, and to their growth and that of the territory, the constitution of the Ecomuseum is finalized (Fig. 12.2).

4 and 5 The two pilot actions in Croatia focused on the development of a brand concept from AZRRI and Zadar County. The accessibility of the Istria pilot area, influenced by its proximity to urban areas like Pula and Rijeka, contributes to a distinct socio-economic trend and robust tourism vitality. Istria County represents a dynamic example of balancing urban development with the preservation of natural and cultural heritage. Moving across the Adriatic, Zadar represents an urban center in Croatia with historical significance and vibrant cultural life. Unlike the Italian regions discussed, Zadar stands as a contrast in terms of accessibility and distance from urban areas. The city's proximity to the coast enhances its tourism vitality and socio-economic trends, showcasing the interplay between urban development and cultural heritage. The virtual Ecomuseum Project of Zadar County pilot area. Both pilot actions aim to develop and implement an innovative territorial management system and create a platform that enables a high level of independence in developing a unique tourist experience. The search for the fifth element has as a mission "To provide each visitor a unique tourist experience, which revives the past, enjoys the present and looks forward to the future" through a unique tourist quest for the 5th element ": fire, water, earth, air. The objective was to develop a tourist attraction brand for Zadar County. AZRRI realized the brand "Sense Route of Istria Heritage" with the involvement of three cities and two municipalities. The Brand will enrich

and fulfill the Istria County inland area, create ground for future itineraries and routes, and attract visitors/tourists to visit and explore the rural areas to lower the coastal pressure.

The pilot actions were tested through a survey. Five different questionnaires were developed for five different action plans. The surveys were conducted by the project partners from January until May 2022. The methodology consisted of questionnaires in an online format and distributed to different groups affected by the project: visitors, tourists, residents/managers from the coastal region, and residents/managers from the hinterland. One of the main results was improved accessibility to inland areas. The majority of participants found pilot actions enjoyable, and the overall assessment of the pilot actions was very high. The respondents also agreed the overall experience enriched their knowledge of the natural and cultural heritage of the region they visited. Pilot action plans were perceived as very positive by the participants, most of them educated, aged between 40 and 50, who look for culture, food, and wine.

Therefore, the project identified and shared innovative models of management of fragile inner landscapes that unleash the potential of their natural assets and cultural heritage in terms of economic development while contributing to their conservation and protection.

Given that sustainable tourism is at the core of a multi-sectoral strategy, such models will be integrated with specific interventions for fostering, and connecting with coastal areas, activities based upon natural and cultural resources, through cooperation and improvement of touristic accessibility. Then, the integrated models will be applied and tested through pilot actions in the project areas to feed the project partners with place-based results and data. The project's main result will be a common action plan, specifying what will be done in each project region to ensure that the lessons learnt from the cooperation project are implemented. This action plan will be developed within an interregional strategy for the sustainable economic and territorial development of inland areas by promoting natural and cultural heritage. The strategy, which will be adopted by the regional authorities involved in the project, will stimulate a cooperation network between different neighboring territories, authorities, and communities/stakeholders, moving from site-specific policies to integrated multi-sectoral policies. Moreover, it will aim to promote each project area as a unique context, where hinterlands are jointly managed in coordination with the related coastal areas.

To Enhance the protection and valorization of the natural and cultural unexploited capital of inland areas by means of tourism development and improved accessibility which ensure the increasing of the specific value of hinterland cultural and natural heritage and their conservation through the empowerment of local communities."

12.2.3 Boost5

The last presented initiative is ***BOOST5—Leveraging results of 5 IT-HR projects to boost touristic valorisation of cultural, off-road, industrial and natural heritage*** (hereafter Boost5)[3] that is a project developed as long-term outcome of Made in-Land. Indeed, the establishment of the Permanent Cross-Border Observatory (PCO) was a key initiative of Made in-Land. As an expected result of the Permanent Cross-Border Strategy, Marche Region has promoted the generation of new projects: the Interreg Italy–Croatia BOOST5, lasting from June 2022 to June 2023. Marche Region participated in the project with seven other Italian and Croatian partners. The BOOST5 project tackles one of the main challenges of the Italian and Croatian Adriatic regions: valorizing and preserving the rich and varied cultural heritage and taking advantage of it in a sustainable way to support economic development. Also, the BOOST5 project aims to overcome the negative impact of a pandemic; the latter proved tourism to be an economic force and a source of livelihood, raising awareness of this potential at the highest levels of government and among the public at large. However, it has also shown the fragility of the sector, which is highly fragmented, imbalanced, and still with low digital intensity. The BOOST5 project capitalizes on the results of five previous IT-HR Interreg Italy–Croatia projects: ATLAS, MADE IN-LAND, UNDERWATERMUSE, TEMPUS, and REVIVAL.

The primary goal of the project was the protection and valorization of the diverse cultural heritage of the Adriatic regions of Italy and Croatia, leveraging this heritage sustainably for economic growth. Additionally, BOOST5 seeks to mitigate the adverse effects of the pandemic on tourism, a sector recognized as both a vital economic engine and a vulnerable, fragmented industry with limited digital integration.

The project was based on the idea that the landscape is a dynamic interplay of natural and cultural networks and their interconnectedness. It recognizes the landscape as a unifying element across the areas and communities involved in the project. Within the Italian and Croatian inland regions, the promotion of novel experiences in gastronomy, sports, arts, and wellness can sustainably enhance the natural and cultural resources, along with the deep-seated traditions of these areas. For this reason, a collaborative planning experience and the best-practice sharing experience are opportunities developed by this project that have boosted positive dynamics in the involved areas.

---

[3] "BOOST 5"—Leveraging results of 5 IT-HR projects to boost touristic valorisation of cultural, off-road, industrial and natural heritage.

Project partnership: TECNOPOLIS Science and Technology Park (LP);; Friuli Innovation Consortium; Sipro Development Agency-Ferrara; Step RI Science and Technology Park of the University of Rijeka Ltd; Marhe Region (and University of Camerino as external consultant); Zadar County Development Agency Zadra Nova; City of Rijeka; ERPAC - Regional Institute for the Cultural Heritage of Autonomous Region of Friuli Venezia Giulia.

Scientific responsible for the Unicam Research Unit: Prof. Massimo Sargolini.

https://programming14-20.italy-croatia.eu/web/boost5.

As a result of the overall analysis and observed dynamics, the project strategy underscores the need for integrated management of the natural and cultural assets, aiming to support mindful, off-season tourism that fosters local economies and virtuous cycles, enhancing the interaction of natural and cultural systems with their surroundings. The idea is to stimulate a coordinated socio-economic development with a conscious management of resources and its territorial heritage.

These objectives are further pursued through the BOOST5 systemic approach on a cross-border scale. The proposed approach helps to overcome the mere local perspective, to enlarge the overall vision, and to create flourishing networks. Moreover, the approach involves both private and public entities in adopting shared strategies for enriching cultural and landscape identities, with the aim to:

- remove major barriers to the full development of sustainable tourism;
- encourage community-led initiatives and new market opportunities through training in innovative, sustainable, and managerial practices;
- foster synergies among various stakeholders and regions to expedite the effective implementation of sustainable and environmentally conscious policies;
- lay the groundwork for future strategic advancements, such as obtaining green accreditations or creating zero-impact tourism labels, and capitalize on upcoming programming period opportunities.

In conclusion, the BOOST5 project, by capitalizing on the successes of the Made in-Land project, reaffirms a shared vision for improving inland areas. This vision integrates landscape enhancement policies, diverse mobility projects, digital communication tools, and a variety of stakeholders in the tourism sector. It solidifies a cross-border strategy that benefits the Italian and Croatian inland areas.

12.3 Discussion and Conclusion

The chapter has mainly presented the goals and the results of three Interreg projects focusing on areas that need to valorize their material and immaterial heritage to boost contextual socio-economic development. After having discussed all the three projects, some conclusions can be drawn. In this perspective, where coherent, the conclusions will show the potential contribution of the projects to the 2030 Agenda's SDGs and their Targets.

In particular, building on the evidence gathered in the projects, this chapter elucidates the significance of data-driven policymaking in the realm of sustainable tourism development. Using data not only assists in identifying and understanding specific challenges within this sector but also plays a crucial role in shaping and directing decision-making processes. Indeed, the literature has already discussed the importance of combining data and indicators to provide a tailored and place-based policymaking process (Battaglia and Annesi 2018). Through a meticulous analysis of this evidence, the chapter aims to illustrate the development of projects that employ a systemic approach and standardized methodologies to enhance natural and cultural aspects. This pivotal approach has addressed several key objectives such as:

- Eliminating barriers to sustainable tourism: the projects gave evidence of how data-driven strategies can help identify and overcome major obstacles that hinder the full development of sustainable tourism offerings. This involves understanding the nuances of local tourism dynamics and addressing issues such as environmental impact, over-tourism, and the preservation of cultural heritage (Sargolini et al. 2022). This dimension is strategic in terms of local contribution to those global priorities that have been discussed in the 2030 Agenda. In this sense, in terms of practical implications, the discussed projects designed an opportunity for the involved areas to develop a sustainable long-term tourism strategy, as institutional implication that should be valorized by the involved areas in their institutional documents and local strategies to promote the development of effective, accountable and transparent institutions at all levels (Target 16), and to institutionalize a gradual contribution to the SDG 8. Somehow, the projects have shaped local dynamics to answer to the development line enclosed within Target 8.8.
- Fostering community-driven initiatives and new market opportunities: an emphasis is placed on engaging local communities in the development process. Capacity-building plans focusing on innovative approaches, sustainability, and managerial skills are crucial for empowering local stakeholders. This empowerment not only aids in creating sustainable tourism products but also opens up new market opportunities that align with the principles of sustainability and cultural authenticity. The full integration of the local community into decision-making processes is also part of the global strategy for sustainable development. Indeed, the 2030 Agenda suggests the implementation of participative, responsive, inclusive, and representative decision-making at all levels (Target 16.7).
- Promoting synergies among diverse stakeholders: the projects advocate the importance of creating synergies between various stakeholders, including local governments, tourism operators, cultural institutions, and environmental groups. Such collaborations are essential for the swift and successful implementation of policies that enhance sustainable accessibility, environmental awareness, and cultural preservation. Fostering a collaborative environment allows stakeholders to share resources, knowledge, and best practices, leading to more cohesive and effective sustainability strategies. The promotion of synergies among diverse stakeholders, moreover, fully responds to the call for action written by the UN in the 2030 Agenda, where they ask governments, parliaments, international, national, and local institutions, as well as people of the civil society, the private sector, and the academy to join the "sustainability journey" in a coordinated manner. In this sense, the Interreg project can be read as a practical tool to fully implement this institutional request (Annesi et al. 2021). Moreover, the participative environment resulting from these projects can contribute to enhance inclusive and sustainable urbanization to ensure the integration of sustainable thinking within human settlement planning and management (Target 11.3).

- Paving the way for strategic advancements: looking toward the future, the projects show how these foundational steps set the stage for more advanced sustainability goals, such as increasing sustainable tourism options, obtaining green accreditations or developing zero-impact tourism labels. These advancements are not only beneficial in terms of environmental and cultural preservation but also serve as a competitive advantage in the increasingly eco-conscious tourism market. In this perspective, data-driven projects and policymaking can support the design, implementation, and monitoring of implemented actions to check if the adopted strategies are contributing to the local development in the proper way. The need to make the settlements more inclusive, safe, resilient, and sustainable (as suggested by the 2030 Agenda in SDG 11) cannot be the result of a not-planned design but should, and have to, be supported by data and indicators to inform policymakers about their choices and citizens about ongoing dynamics.
- Leveraging future opportunities: finally, the projects evidence how these initiatives can be built upon in upcoming programming periods. This involves adapting to evolving market trends, technological advancements, and changes in environmental policy. By staying ahead of these trends, regions can ensure that their tourism offerings remain sustainable, culturally relevant, and appealing to a broad audience. In this perspective, the regional approach to this topic can support the positive economic, social, and environmental links between urban, peri-urban, and rural areas by strengthening national and regional development planning (Target 11. a).

In summary, this chapter demonstrates the transformative power of data-driven policymaking in sustainable tourism development. By adopting a holistic approach that values natural and cultural enhancement, community involvement, and strategic collaboration, regions can pave the way for a more sustainable and prosperous future in the tourism sector.

References

Annesi N, Battaglia M, Gragnani P, Iraldo F (2021) Integrating the 2030 agenda at the municipal level: multilevel pressures and institutional shift. Land Use Policy 105:105424

Barca F (2009) Agenda for a reformed cohesion policy. European Communities, Brussels

Barca F, Casavola P, Lucatelli S (2014) *Strategia nazionale per le aree interne: Definizione, obiettivi, strumenti e governance*. Ministero dello Sviluppo Economico, Dipartimento per lo Sviluppo e la Coesione Economica, Unità di Valutazione degli Investimenti Pubblici

Battaglia M, Annesi N (2018) Efficient and nice–urban sustainability and quality of life: the socioeconomic perspective. In: Cocci Grifoni R, D'Onofrio R, Sargolini M (ed) Quality of life in urban landscapes: in search of a decision support system, pp 141–149

Battaglia M, Annesi N, Pierantoni I, Sargolini M (2019) Future perspectives of sustainable development: an innovative planning approach to inner areas. Experience of an Italian alpine region. Futures 114:102468. ISSN: 0016-3287. https://doi.org/10.1016/j.futures.2019.102468

Canale RR, De Simone E, Di Maio A, Parenti B (2019) UNESCO World Heritage sites and tourism attractiveness: the case of Italian provinces. Land Use Policy 85:114–120. https://doi.org/10.1016/j.landusepol.2019.03.037

Cetara L, Pierantoni I (2014) Landscapes, economies and local communities. In: Gambino R, Sargolini M (ed) Mountain landscapes. A decision support system for the accessibility. List Lab Publisher, Trento. ISBN: 9788898774388

Dastgerdi AS, Pierantoni I, Sargolini M, Stimilli F (2020) Toward an innovative strategic approach for sustainable management of natural protected areas in Italy. Geogr, Environ, Sustain 68–75, ISSN: 2071–9388. https://doi.org/10.24057/2071-9388-2019-143

European Commission: Directorate-General for Regional and Urban Policy (2007) Cohesion policy 2007-13: commentaries and official texts. European Regional Development Fund. European Social Fund. Cohesion Fund. European Grouping of Territorial Cooperation. Instrument for Pre-Accession Assistance. Publications Office of the European Union. https://op.europa.eu/en/publication-detail/-/publication/debf5b93-194a-4fe2-8bcb-fb630a0d12d2

Leask A (2016) Visitor attraction management: a critical review of research 2009–2014. Tour Manag 57–2016:334–361. https://doi.org/10.1016/j.tourman.2016.06.015

Pierantoni I, Sargolini M (2020) Protected areas and local communities: a challenge for inland development. ListLab Publisher, Trento. ISBN: 9788832080414

Sargolini M, Gambino R (2014) Mountain landscapes. A decision support system for the accessibility. List Lab Publisher, Trento. ISBN: 9788898774388

Sargolini M, Shirvani Dastgerdi A, Broussard Allred S, Chatrchyan A, De Luca G (2020) Climate change and sustaining heritage resources: a framework for boosting cultural and natural heritage conservation in central Italy. Climate 8(2):26. https://doi.org/10.3390/cli8020026

Sargolini M, Pierantoni I, Polci V, Stimilli F (eds) (2022) Progetto Centro Italia. Nuovi sentieri di sviluppo per l'Appennino Centrale interessato dal sisma del 2016. Pescara, Carsa Editions. ISBN: 978-88-501-0411-6

De Siano R, Canale RR (2022) Controversial effects of tourism on economic growth: a spatial analysis on Italian provincial data. Land Use Policy 117:106081. https://doi.org/10.1016/j.landusepol.2022.106081

Author Index

A
Annesi, Nora, 239
Antić, Marija, 61

B
Branković, Milena Dinić, 205

C
Čolig, Marina, 33

D
Đekić, Jelena, 205
Đorđević, Dejan Ž., 61, 85
Drezgić, Saša, 33
Đurkin, Danica, 61
Džunić, Marija, 13

G
Garau, Chiara, 173

I
Igić, Milica, 205

J
Jakšić, Milena, 103
Jovanović, Sonja, 153

K
Kregar, Kristina, 33
Krstić Srejović, Ana, 103

L
Langović, Milica, 85
Lepojević, Vinko, 153
Lu, Chaoru, 189
Luè, Alessandro, 1

M
Marinković, Srđan, 13
Marjanović, Ivana, 13
Milić, Vesna Janković, 153
Mimović, Predrag, 103
Murgante, Beniamino, 173

O
Obradović, Tanja, 205
Östh, John, 189

P
Piatkowski, Daniel, 189
Pierantoni, Ilenia, 239

R
Rađenović, Tamara, 205

J. J. Stanković et al. (eds.), *Regional and Urban Challenges and Opportunities for Data-Policy Interactions*, The Urban Book Series,
https://doi.org/10.1007/978-3-032-08592-4

S
Šantić, Danica, 85
Sargolini, Massimo, 239
Simić, Ivana, 131
Stanković, Jelena J., 1, 13

T
Tan, Wendy, 173

V
Van der Laag, Claudia, 1, 173

MIX
Papier aus verantwortungsvollen Quellen
Paper from responsible sources
FSC® C105338

If you have any concerns about our products,
you can contact us on
ProductSafety@springernature.com

In case Publisher is established outside the EU,
the EU authorized representative is:
Springer Nature Customer Service Center GmbH
Europaplatz 3, 69115 Heidelberg, Germany

Printed by Libri Plureos GmbH
in Hamburg, Germany